Acknowledgments

We, the authors, would like to thank the people who helped make this book possible. We especially want to thank the staff of the Pacific Northwest National Laboratory's Molecular Biosciences Department who graciously allowed their work to be an example in this document. In particular, special thanks to Ron Thomas who took the time to review our summary, Dr. Kevin Groch, the assessment team leader, and the remainder of the assessment team. We also want to thank those who took their valuable time to review the document and provide us with comments: Mark Boylan of WASTREN, Ernie Kuhn of Washington State University, and Sandra Cannon, Tapio Kuusinen, Keith Peterson, Elizabeth Raney, and Jim Weber of Battelle. Ultimately, any remaining errors or idiosyncrasies remain ours, but their help in improving our text as a whole is appreciated. Finally, we would like to thank Michael Perkins and his excellent work preparing the graphics.

Pollution Prevention Opportunity Assessments

FOR
RESEARCH AND DEVELOPMENT LABORATORIES

JILL ENGEL-COX
KIM FOWLER

PACIFIC NORTHWEST NATIONAL LABORATORY

Battelle Press

Disclaimer

This document was prepared by the Pacific Northwest National Laboratory operated by Battelle Memorial Institute for the U.S. Department of Energy under Contract DE-AC06-76RLO 1830. Because it is not possible to envision the conditions under which this information will be used, neither the authors, nor Battelle, nor the Department of Energy MAKES ANY WARRANTY OR REPRESENTATION, EXPRESSED OR IMPLIED, that use of the methods discussed herein will be successful in complying with local, state, or federal laws and regulations, or that such would be safe. Those using this information must assure themselves that they are operating safely and within the law. This document is intended to serve as a resource, but not as a substitute for legal advice or good judgment.

Library of Congress Cataloging-in-Publication Data

Engel-Cox, Jill.
Pollution prevention opportunity assessments for research & development laboratories / by Jill Engel-Cox and Kim Fowler.
p. cm.
Includes index.
ISBN 1-57477-070-5 (alk. paper)
1. Laboratories—Waste minimization. 2. Pollution prevention.
I. Fowler, Kim, 1965-
TD899.L32E53 1999
628.4—dc21 99-14313
CIP

Printed in the United States of America

Battelle Press
505 King Avenue
Columbus, Ohio 43201-2693
614-424-6393; 1-800-451-3543
Fax: 614-424-3819
Homepage: http://www.battelle.org/bookstore
E-mail: press@battelle.org

Table of Contents

Introduction

Chapter 1

Preventing Pollution in the World of Research and Development

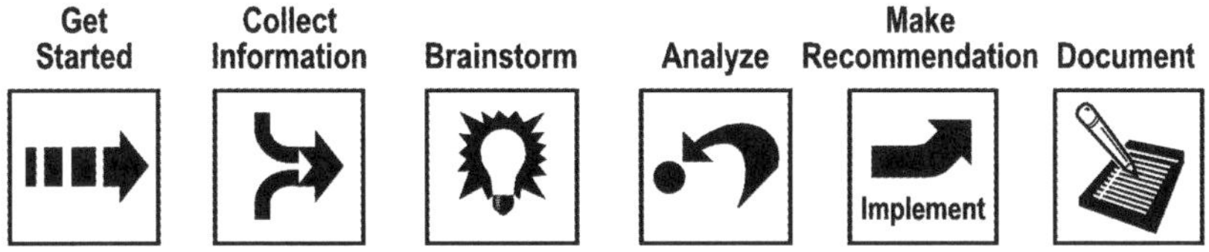

1. *Choose a diverse team, including a researcher to lead the assessment.*
2. *Gather data about key waste-generating activities by identifying the material usage, quantifying waste streams, and walking through facilities and laboratory space where activities are being conducted.*
3. *Brainstorm pollution prevention opportunities.*
4. *Research and analyze pollution prevention opportunities for waste reduction, cost savings, and return on investment.*
5. *Make recommendations for pollution prevention implementation projects based on the waste and cost analysis.*
6. *Document your work and implement the opportunities.*

INTRODUCTION

Reducing waste, material use, and cost in a research and development (R&D) environment requires a new approach to the standard pollution prevention tools developed for production or construction activities. Unlike most staff involved in manufacturing processes, research scientists and engineers use a large variety of chemicals and materials resulting in many small waste streams. Moreover, R&D waste streams can often change, as research requires different and ever-diverging directions of work. Yet, despite the fact that these are small waste streams, as a whole they can result in large amounts of very hazardous and complex materials. Waste from R&D laboratories can collectively span the entire spectrum of waste (from flammable to toxic and from liquid to solid) and be regulated by a vast array of local and federal statutes. The existence of many different types of chemicals and wastes requires an infrastructure of hazard communication, emergency response, waste management, and storage far beyond many manufacturing firms with larger quantities but fewer types of waste.

Until now, most of the information on how to prevent pollution has been directed at manufacturing and construction. This handbook helps to fill that gap by presenting demonstrated tools and examples of ways to reduce waste generated by R&D processes, projects, and programs. It allows researchers, environmental compliance staff, and managers of R&D programs in private industry, government, and university research departments to identify ways to reduce chemical and material use, to minimize the generation of waste, and to lower the associated costs. This handbook is designed to be easily adapted to the requirements of different laboratories and programs.

The Pacific Northwest National Laboratory (PNNL) uses the *pollution prevention opportunity assessment* method as the primary tool for identifying ways to reduce pollution and avoid costs. The Environmental Protection Agency originally developed the pollution prevention opportunity assessment method for manufacturers, calling them "waste minimization opportunity assessments." The Department of Energy later called its first version of the Environmental Protection Agency method "process waste assessments." However, at PNNL, we have modified the assessment approach to suit the conditions found in research laboratories. The assessment is a simple and systematic method to find pollution prevention opportunities, take advantage of them, and document the efforts.

All of the techniques and assessments detailed in this handbook were implemented and tested at PNNL. PNNL has approximately 3,500 research scientists and support staff working on over 2,000 projects, primarily focused on environmental and energy research and development. In this R&D environment, numerous opportunities are available for developing and testing pollution prevention strategies. The pollution prevention opportunity assessment process is continually adapted to meet the needs of the researchers while accomplishing the goal of preventing pollution.

This introductory chapter covers some basic concepts, such as describing pollution prevention, discussing what a pollution prevention opportunity assessment is, noting the benefits of assessments, and showing how they fit into a pollution prevention program. Each subsequent chapter covers a step in the pollution prevention opportunity assessment process. At the beginning of each chapter is a list of the activities to complete that step. Each chapter also includes an example of how a team of researchers completed that step and the results. A detailed report from the example is included in the appendix.

Pollution Prevention Techniques

The hierarchy of source reduction, reuse/recycle, treatment, and disposal is an easy way to categorize some of the common pollution prevention techniques used in laboratories.

Source Reduction

Reduce the scale of the experiment: Reduce the sample size and or perform the experiment using microscale techniques or technology.
Substitute materials: Replace a hazardous substance with a nonhazardous substance.
Go electronic: Replace a process that historically was performed with chemicals with one that can be performed on a computer.
Modernize equipment: Review the equipment catalogs to see if the up-to-date versions of your equipment use less material and/or generate less waste.
Redesign the experiment: Modify the procedures of the experiment to reduce the number of steps and to incorporate other source reduction opportunities.
Manage inventories: Order only those chemicals needed to perform the experiment in the quantity needed now; that is, do not order in bulk just because the chemical purchasing cost appears to be less expensive.
Segregate materials: Keep hazardous and nonhazardous materials separate from the moment they arrive in the laboratory.
Practice good housekeeping: Keep the laboratory space clean, redistribute or dispose of materials when they are no longer needed, maintain equipment so that it is performing at its optimal level.

Reuse/Recycle

Close the loop: Modify the process so that the waste outputs can be used as inputs.
Exchange materials: Communicate with other researchers about the availability and/or need for specific chemicals and laboratory materials so that excess materials can be used by others instead of being disposed of as waste.
Use reusable laboratory equipment: Use laboratory glass and equipment that can be cleaned and reused versus disposable materials wherever possible.
Recycle the recyclables: Where materials cannot be reused, coordinate with local recyclers or vendors for a recycling program that accepts laboratory materials.
Purchase supplies that are recyclable and that contain recycled content: Items that contain recycled content should be purchased wherever possible so that a continued market for recycling exists. If one-time use is required, purchase items that can be recycled.

Treatment

Neutralize: Adjust the pH of a liquid waste stream so that it is no longer considered hazardous and can be disposed into the sewer. Regulatory approval is required for on-site treatment of hazardous waste.
Evaporate: Remove the water from the waste stream prior to disposal. This technique reduces volume but often increases the toxicity of the now concentrated waste. Regulatory approval is required for on-site treatment of hazardous waste.
Compact: Reduce the volume of the waste by crushing it into a smaller volume prior to disposal.
Recover waste: Turn end-of-the-pipe waste into a saleable product.

Disposal

Document: Carefully record all of the constituents of the materials you will dispose of to ensure proper characterization and disposal.
Coordinate with waste management staff: Submit your waste and all documentation to waste management personnel and ask them any questions you may have about storage, documentation, or handling.

What is Pollution Prevention?

The first approach to preventing pollution is *source reduction*—reducing waste at the source before it is generated. This is the easiest and most cost-effective way to eliminate or at least reduce waste streams, since nothing has to be treated, boxed, or barreled. Source reduction may mean replacing a waste-producing process with a cleaner process, such as employing microscale chemistry methods or alternative nonhazardous chemicals. Or it may mean finding ways to do work electronically instead of with chemicals, such as using digital cameras instead of film-developing processes.

The second approach is *reuse and recycle*—reusing chemicals and materials or reprocessing instead of disposing of them. Reusing or redistributing chemicals and materials within a laboratory can save thousands of dollars in purchase and disposal costs, as well as facilitate good chemical management. Recycling can be done within the laboratory, such as with solvent distillation units for more consistent chemical waste streams, or through contracts with outside vendors, such as for oil re-refining or battery regeneration. Often, these recycling techniques and contracts can be done without additional costs since the material being recycled has value—one person's waste is another's gold.

When source reduction and reusing and recycling are not feasible, *treatment* is an option—reducing the toxicity or volume of waste before safely disposing of it. Neutralizing chemicals is one direct way of accomplishing this. Considerable savings in disposal fees (and in laboratory space) have been possible after hazardous waste is made suitable for process sewer disposal. Often, treatment can be done on-site with common laboratory equipment and processes, once regulatory approval has been received.

What is a Pollution Prevention Opportunity Assessment?

A *pollution prevention opportunity assessment* is a systematic, documented approach to determining where pollution prevention opportunities exist. Once opportunities are identified, the assessment can also reveal the best means of implementing the opportunity. Far from being a one-time housecleaning guide, the pollution prevention opportunity assessment maps out future pollution prevention activities and savings opportunities so that the benefits can keep on being generated.

A pollution prevention opportunity assessment also helps determine the cost of the present waste streams, the cost of changing to a cleaner process, the cost savings of a cleaner system, and the time needed to repay your investment. Determining the cost impacts of waste streams is an essential part of the pollution prevention process. The assessment can point toward innovative or emerging technologies that will be cleaner, safer, and more cost-effective, and will sometimes yield better scientific results.

The steps to completing an assessment are basic but effective (see Figure 1). They are detailed in the chapters of this handbook:

1. Choose a diverse team, including a researcher to lead the assessment.
2. Gather data about key waste-generating activities by identifying the material usage, quantifying waste streams, and walking through facilities and laboratory space where activities are being conducted.
3. Brainstorm pollution prevention opportunities.
4. Research and analyze pollution prevention opportunities for waste reduction, cost savings, and return on investment.

5. Make recommendations for pollution prevention implementation projects based on the waste and cost analysis.
6. Document your work and implement the opportunities.

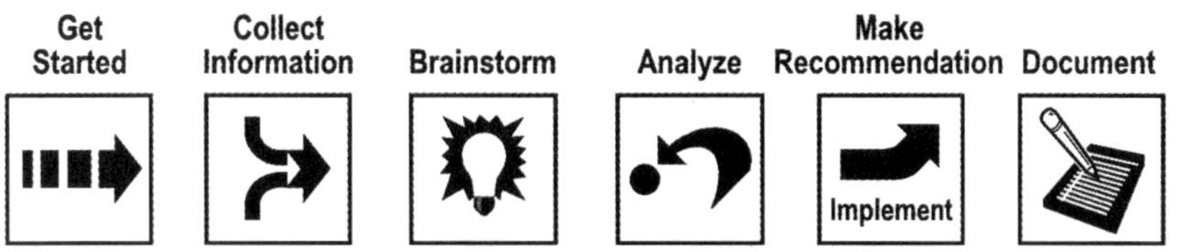

Figure1: Steps in a Pollution Prevention Opportunity Assessment

Why Do a Pollution Prevention Opportunity Assessment?

The benefits from including pollution prevention in R&D laboratory operations have been well documented. A pollution prevention opportunity assessment results in recommendations to reduce waste, and if these are implemented, the benefits can include

- *Cost savings*, including reduced material purchase costs, reduced chemical management and waste disposal costs, reduced regulatory requirements and compliance costs, and increased data quality and work efficiency
- *Safety improvements and risk reduction*, including reduced staff exposure to chemicals and reduced potential for accidents and spills
- *Process efficiency*, including more efficient use of raw materials and better design of experiments and research projects
- *Environmental stewardship*, including reduced potential for releases to the environment, reduced raw material and energy consumption, reduced waste generation, and reduced ecological degradation
- *Good public relations*, including reduced inventories and releases reported, increased positive activities to share with community, and improved relations with regulators
- *Enhanced scientific operations*, including improved quality and quantity of research output by utilizing new technologies or processes.

PNNL has had considerable success with performing pollution prevention opportunity assessments on laboratory operations. Throughout the handbook, a description of one of those successes will be used as an example for each of the steps.

How Do Assessments Fit Into a Bigger Picture?

Pollution prevention opportunity assessments are powerful tools for learning more about laboratory activities and determining ways to reduce waste and save money. The intent of performing an assessment is to identify which pollution prevention opportunities should be implemented and which are not beneficial or cost-effective. Without implementation, a pollution prevention opportunity assessment is merely another paper study and no waste will be reduced. Pollution prevention opportunity assessments are most effective when conducted in the context of a pollution prevention program that provides support and encouragement to researchers in reducing their waste streams.

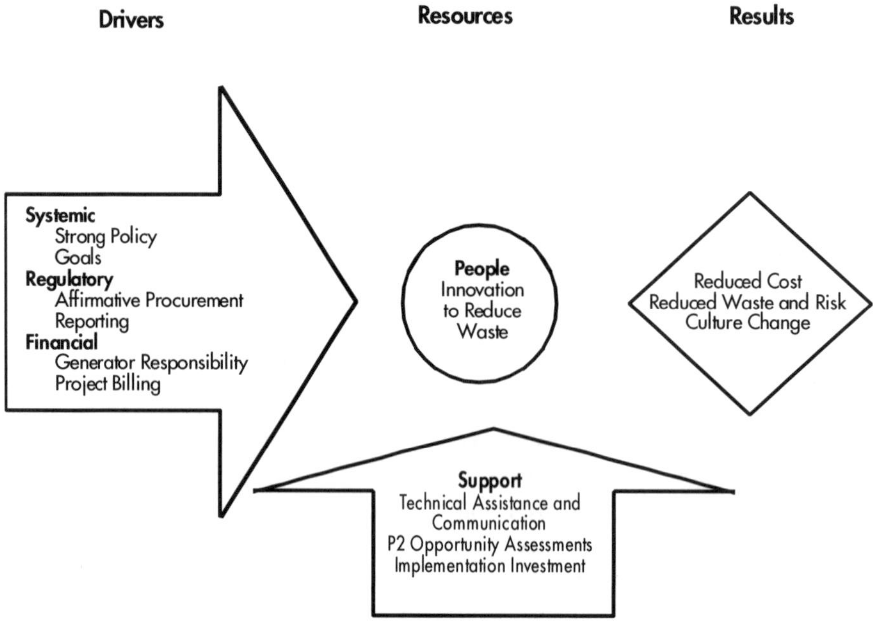

Figure 2: Integrated Pollution Prevention Program

At PNNL, pollution prevention opportunity assessments are part of a comprehensive pollution prevention program, as shown in Figure 2. This program includes financial and institutional incentives that encourage staff to initiate pollution prevention activities, including management support and a cost chargeback system. It also provides the technical support to help researchers find the resources and information to implement the opportunities recommended by their assessments and recognizes them for their positive work. In turn, pollution prevention opportunity assessments are critical to achieving larger goals for a pollution prevention program, especially helping foster a culture change among researchers and staff, allowing them to think about their work in terms of its impact on the environment.

An Overall Pollution Prevention Philosophy

The key to conducting assessments and reducing waste in an R&D environment is to remember that the laboratories, offices, and classrooms are filled with intelligent and creative people who care about their work and its potential impact on the environment. When given funding for work time, technical assistance, a basic assessment methodology, and the freedom to review their own activities, research scientists and engineers will develop creative solutions to reduce the waste they create and the materials they use. We hope this handbook will help you successfully prevent pollution in your laboratory.

How to Use This Handbook

Chapters 2 through 7 contain step-by-step instructions on how to perform a pollution prevention opportunity assessment in your laboratory. The instructions include a short list of tasks, a descriptive discussion on how to perform the tasks, and an illustrative example for each of the steps. Chapter 8 summarizes the whole process, and the appendix provides the complete final pollution prevention opportunity assessment report for the example used throughout Chapters 2 through 7. The handbook should be used as a guide to conducting an assessment of your own laboratory and be considered flexible enough to meet your own needs and resources.

Background Information for the Example

In 1995, the 331 Building at the Pacific Northwest National Laboratory received compliance violations for mishandling waste and hazardous materials that resulted in a temporary shutdown of all research activities in the building. As a result, the Molecular Biosciences Department, one of the occupants of the 331 Building, decided to review its waste-generating activities for pollution prevention opportunities. Ultimately, it succeeded in significantly reducing its waste streams, improving its operations and quality of research, receiving praise from regulators, and becoming a model laboratory for continuous environmental improvement.

The Molecular Biosciences Department consists of 17 research staff conducting modern biological research, including molecular biology, structural biology, and biochemistry through whole animal and toxicological studies. Their work is the integration of multiple small-scale processes that are performed intermittently, using a number of reagents. The individual processes produce waste in the range of a few hundred grams to several kilograms per year. The wastes generated by a number of small processes are collected for storage in a Satellite Accumulation Area.

The Molecular Biosciences Department staff were already aware of some of the possibilities in pollution prevention, even before deciding to conduct a pollution prevention opportunity assessment. The department had purchased a Molecular Dynamics FluorImager to minimize the use of phosphorus-32 (a radioactive tracer) and reduce photo-processing needs in its research laboratories. Phosphorus-32 had generated 100 to 150 L of low-level radioactive waste annually in the Molecular Biosciences Department laboratories. By using the FluorImager, the department no longer needed a radiation technician, saving $100,000 annually.

To begin to improve its process, the Molecular Biosciences Department established a goal in its revised Environment, Safety, and Health Plan: "We will continue to seek waste minimization opportunities and support the PNNL pollution prevention program by reducing the amount of hazardous materials that may eventually become waste, and by encouraging reuse and recycling to the greatest extent economically practicable." The outcomes the department hoped to achieve through performing the pollution prevention opportunity assessment included

- Reduced liability of future compliance problems
- Improved relations with regulators
- Reduced waste
- Improved working conditions and worker morale
- Reduced cost.

See the end of each chapter for the continuation of this example. The experiences of the Molecular Biosciences Department are used to describe each of the pollution prevention opportunity assessment steps.

Get Started

Chapter 2

Focusing on the Environment

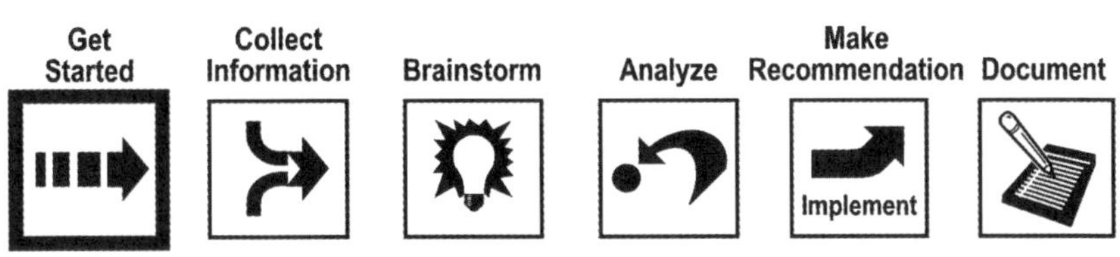

1. *Establish the scope of the assessment.*
2. *Secure financial resources.*
3. *Gain management support.*
4. *Identify the research team leader and the assessment team.*
5. *Assign a pollution prevention specialist.*
6. *Set a schedule and kick off the assessment.*

Get Started

In the early stages of assessments, team members often encounter resistance from their fellow research laboratory staff: "We don't generate that much waste" or "We are already doing everything we can." Typically, they may have made some changes to their research processes, picking the opportunities that are easy to implement—the low-hanging fruit. However, by the time the assessments have ended, these same researchers are writing proposals and approaching management in order to implement new pollution prevention activities. The assessment process can uncover a range of new possibilities that prevent laboratory waste, save money, and improve research. New technologies and processes are always being developed, therefore finding alternatives that both reduce waste and enhance the scientific research requires a commitment to continuous improvement.

The early steps of establishing a scope, securing funding, getting management support, forming a team, and setting up the assessment are essential to the laboratory's success at implementing ways to reduce waste.

Establish the Scope of the Assessment

For R&D pollution prevention opportunity assessments, the scope or the boundaries of the assessment should be structured to fit the needs of those performing the assessment. The scope is those activities and/or locations that the assessment will review. The scope can cover a single waste stream, a single research activity or process, a set of waste streams or processes, all the activities within an organization, or any combination of these. How the scope is set can depend on the nature of the waste streams, nature of research projects, physical location of laboratories, number of research staff involved, or all of the above.

Examples of pollution prevention opportunity assessment scope include

- A department, technical group, or set of laboratories located in the same building
- A group of similar waste streams, research processes, or projects
- Waste streams from similar equipment used on various projects (for example, mass spectrometers, vapor degreasers)
- One-time, high-volume waste projects that are occurring currently or in the near future (for example, laboratory clean-outs, moves, new construction).

Ideally, the scope is not so large that more than 10 pollution prevention opportunities result from the assessment, or else the assessment may become too big and lose focus. An exception to this would be when most of the opportunities identified are easy to implement. If the assessment scope seems too big for one team to handle in a reasonable time, the assessment can be split into two and conducted separately with different teams.

Secure Financial Resources

Researchers need time to conduct the assessments. Thus, they require funding for their time. Depending on the scope and the level of detail, an assessment can require from 60 to 100 staff hours. Several sources can be tapped for funds to conduct an assessment. First, the management of the staff covered in the assessment's scope may be willing to fund the team from internal sources. Since they are the organization that will benefit from any cost-saving opportunities that may result from the assessment, this may be a good investment for them. Another source can be internal environmental, safety, and health funds. Internal compliance organizations often have funds to conduct inspections, and these may sometimes be used for assessments. They may even have a pool of funding set aside just for pollution prevention opportunity assessments. Some universities have employed students enrolled in a pollution prevention, environmental science, or chemistry class to complete assessments as part of a course. This gives students excellent hands-on experience and can be a low-cost way to complete the time-consuming information-gathering and analysis portions of an assessment. When students conduct the assessment, however, care must be taken that an internal commitment to implement the results of the assessment remains with the research staff. Finally, grants or technical assistance for assessments may be available through local or city regulatory agencies.

Gain Management Support

Without management support, an assessment cannot be effective. However, management support can come in several ways and at various levels. The most basic sort of support may be an agreement to conduct an assessment. Other forms of management support may include actually being a part of the assessment team, providing resources to conduct the assessment, providing resources to implement opportunities, and/or making completion of the assessment or implementation of pollution prevention opportunities a part of the group's performance goals. These levels of support feed each other as a group or department increases its depth of knowledge and enthusiasm for pollution prevention. Support may also increase with a group's ability to find resources to implement pollution prevention opportunities.

Critical Elements and Expectations

At PNNL, two program elements are essential for a successful pollution prevention opportunity assessment:

- The assessment team leader is a researcher who works in the area being assessed
- The assessment team is provided with a budget, a general methodology, and as much pollution prevention technical assistance as needed.

The leader and the team have two critical expectations to meet:

- Pollution prevention is the focus of the assessment, i.e., the team looks for opportunities to reduce at the source first, then reuse/recycle, and finally treatment.
- The final product is the completed worksheets (discussed in Chapter 7).

Identify the Research Team Leader and the Assessment Team

The team is the key to a successful pollution prevention opportunity assessment and implementation of pollution prevention projects. The selection of the team leader is very important and should be done first. The leader is often the instigator of the assessment, a researcher who is interested enough to volunteer, or a good research leader who has been chosen by management. Most important, the team leader needs to be a respected member of the research projects or groups that are being assessed. A respected research team leader will provide the incentive and enthusiasm for implementation of the results of the assessment. The team leader identifies key wastes or processes to evaluate, rallies staff to provide data for the assessment, and recommends pollution prevention activities to management. It is up to the team leader to make sure that data are gathered and the pollution prevention opportunities are adequately investigated. The team leader needs to be a good organizer, willing to talk to management to resolve issues and to request implementation funding. The team leader needs to work with a diverse set of staff, such as compliance, waste management, and research staff.

Once selected, the team leader is responsible for forming the team. A team can be formally structured or a loosely associated group of people who participate in portions of the assessment but not all of it. The team can include

- *One to five key researchers in the laboratories being assessed*—Research staff are essential to this process. Researchers are more likely to accept ideas and new ways of working when their colleagues have been involved in identifying pollution prevention opportunities. Teams should also include any researcher who might have to modify research practices significantly as a result of this assessment.
- *A pollution prevention specialist*—A pollution prevention specialist can provide a usable methodology, share pollution prevention technical information gained from past assessments, and facilitate meetings and brainstorming. The pollution prevention specialist often can offer technical expertise in uncovering pollution prevention opportunities that may be hidden even from those who work in the laboratory every day.
- *A management representative*—Managers can provide insight into the intent of practices or policies. Participating in the assessment, even if only in the walkthrough and the brainstorming sessions, helps managers understand and support the team's recommendations and requests for implementation funds. Also, getting managers into laboratories often gives them insight into the working conditions of the researchers and what it takes to get work done.
- *A waste management specialist*—Waste management specialists contribute knowledge of waste streams, including what happens to the waste once the research is completed. This person can also provide assistance in researching the pollution prevention ideas.
- *An environmental compliance specialist*—To keep the team from introducing new ideas that raise compliance issues, an environmental compliance specialist can assist with background and research.

The assessment team also should invite participation from those interested in pollution prevention who could contribute to the scope of the assessment. Four to eight team members seems to work well, but larger teams with more members have also been successful.

Assign a Pollution Prevention Specialist

The pollution prevention specialist who is assigned to the pollution prevention opportunity assessment can provide the level of assistance that the team leader decides is appropriate. The specialist will help keep the focus of the assessment on pollution prevention and will help ensure that the pollution prevention opportunity assessment worksheets are completed when the assessment is done (see Chapter 7).

Often, the pollution prevention specialist offers a wide range of assistance that includes

- Identifying and refining the general scope of the assessment as defined by the team leader
- Encouraging management involvement
- Setting the assessment schedule with the team leader
- Providing background information to assist with the assessment, such as previously performed assessments or waste generation data
- Participating in the laboratory walkthrough
- Facilitating the brainstorming session
- Researching pollution prevention opportunities at the request of the team leader
- Reviewing and assisting with the completion of the worksheets.

The pollution prevention specialist can also be responsible for many of the administrative tasks, such as making sure the appropriate managers are aware that the assessment is being performed, making the final assessment available for future reference, and helping prepare and submit implementation proposals where appropriate. The pollution prevention specialist may also follow up with the team to discover what was or was not implemented and what cost and waste savings actually were achieved.

However, do not expect the pollution prevention specialist to perform the entire assessment or to be the team leader. The assessment needs to be conducted by the research staff, who, as a result, will ensure that, as the benefits become obvious, pollution prevention opportunities will be implemented.

Set a Schedule and Kick Off the Assessment

Once the pollution prevention opportunity assessment team has been named and the scope has been established, it is time to set the schedule and hold an assessment kick-off meeting. Scheduling may be affected by a variety of factors, but always involves working around the ongoing research tasks of the laboratory. Nevertheless, a schedule should follow a simple plan that is clear to team members and laboratory staff. For instance, a week-by-week schedule could look like this:

Week 0.	Management approves and assessment begins.
	Management assigns a team leader or a researcher volunteers to lead the assessment.
Week 1.	Pollution prevention specialist provides a copy of this book, samples of previously performed assessments, and blank worksheets to the team leader.
	Team leader determines the team members and provides introductory materials to team members.
Week 2.	Team leader holds kick-off meeting attended by all team members.
Week 3.	Team completes draft version of Worksheets 1 and 2 (see Chapter 7).
	Team leader distributes completed draft copies of Worksheets 1 and 2 to team.

Week 4. Team performs the walkthrough of laboratory spaces (1 to 2 hours).
Pollution prevention specialist facilitates the brainstorming session (scheduled for at least 2 hours).
Team leader assigns the research and completion of Worksheet 3 to team members.

Week 4–8. Team researches and analyzes opportunities.
Team completes drafts of Worksheet 3.

Week 8. Team submits drafts of Worksheet 3 to the team leader.
Team leader distributes drafts of Worksheet 3 for comments.

Week 9. Team leader prepares drafts of Worksheets 4 and 5.

Week 10. Team leader incorporates Worksheet 3 comments into the draft document.

Week 11. Team leader distributes a complete draft of the completed pollution prevention opportunity assessment to the entire team.

Week 12. Team leader incorporates final comments.

Week 13. Team completes assessment and presents results to management and others.

The kick-off meeting is held with the entire team and is intended to make sure everyone understands the process. It can be short, e.g., 30 to 60 minutes. During the meeting, the pollution prevention specialist provides an overview of the process. The team leader describes the general scope and schedule for the assessment. The leader also asks for suggestions to refine the assessment scope and solicits information on key waste streams, processes, laboratory spaces, and projects on which the team thinks that it should focus its effort.

When the schedule is set and the kick-off has taken place, it is time to begin the assessment!

Example

The PNNL Molecular Biosciences Department decided to limit the scope of its assessment to its own departmental activities within a single building. This excluded other departments working in that building and other Molecular Biosciences Department activities occurring in other buildings. It did not try to focus on certain procedures or activities, instead choosing to do a broad review of work within its influence.

Management was supportive of the assessment because of past compliance difficulties, its initial relatively small success with pollution prevention, its strong commitment to doing the "right thing," and its recognition that waste reduction techniques were also economical. To further drive the assessment, management included a requirement for conducting an assessment in the department's revised Environmental, Safety, and Health Plan. Department management contacted the PNNL pollution prevention team to learn more about pollution prevention opportunity assessments and what resources were available. The pollution prevention team was able to fund its assessment, including the time for both the team leader and the team members and to provide technical assistance.

The department selected one of its research scientists to conduct the assessment. The scientist was knowledgeable about the types of research being performed by the department in the building, as well as about the chemicals and processes involved. Members of the team included

- A researcher assigned as assessment team leader
- A Molecular Biosciences Department manager
- The department's internal Environmental, Safety, and Health Coordinator
- An environmental compliance expert
- A materials and waste management expert who ordered the department's chemicals and shipped its waste
- A pollution prevention specialist.

The team leader met with the pollution prevention specialist and received guidance on how to conduct a pollution prevention opportunity assessment. They set the following schedule:

April 11	Review waste streams, select team, complete planning
April 12	Hold kick-off meeting
April 26	Complete Worksheets 1 and 2
May 3	Walk through facility and hold brainstorming session
May 31	Research opportunities and complete Worksheet 3
June 15	Select and plan opportunities and complete Worksheets 4 and 5
July 15	Complete assessment and begin pursuing implementation funds.

Collect Information

Chapter 3

Raising Your Awareness About the Activity

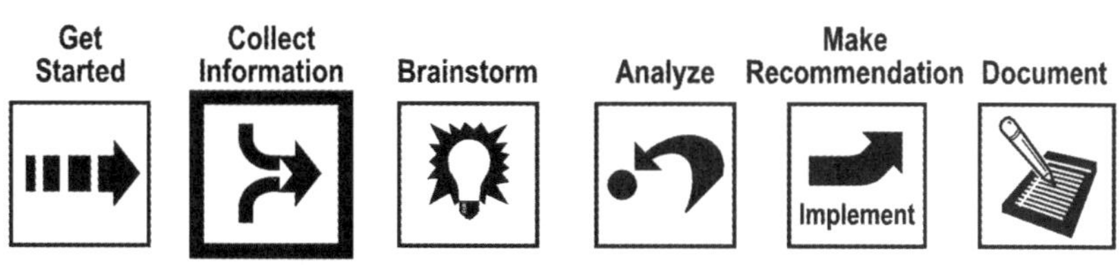

1. *Understand your chemical usage.*
2. *Determine how much waste is being generated.*
3. *Document the scope of the assessment.*
4. *Research similar pollution prevention successes.*
5. *Complete a walkthrough of the assessment area.*

Collect Information

Now that you have committed to performing an assessment, obtained funding, brought together a team, enlisted management support, and set a schedule, the next step is to start collecting and documenting the information needed to perform the assessment.

The process of gathering data is important, but it is only one of the beginning steps. The challenge is to gather enough information to have a good understanding of the activities in the laboratory that generate waste without making the task overwhelming by putting too much emphasis on data collection. No more than one-third of the time spent on the assessment should be spent in the data collection step.

The type of information collected depends on the scope of the assessment; where you find the information depends on the data systems already in existence within your organization.

Understand Your Chemical Usage

This task involves understanding not only which chemicals are being used, but also how they are being used. Many chemicals and processes may be available that are less hazardous or generate less waste volume. However, these substitutes may or may not be appropriate for the type of work being performed. Therefore, this task cannot be performed in a vacuum. It may involve checking chemical purchasing lists and chemical inventories, but it must involve talking with the researchers to understand if special requirements, restrictions, or specifications are associated with their research.

First, gain a general understanding of the chemicals being used. To do this, review chemical inventory lists and chemical purchases. Once you have a general view of the chemicals, talk with researchers to understand how the chemicals are being used. This information can be obtained in a variety of ways, ranging from conducting informal conversations to supplying a survey to each of the researchers. The following sample questions address the reasons for a researcher's use of particular chemicals:

1. Identify the repeatable processes performed in your laboratory space.
2. Identify any upcoming or current one-time processes that are likely to generate a large quantity of waste.
3. What chemicals are needed to perform this process?
4. What kind of and how much waste does this process generate
5. Identify the processes that generate the largest volumes of waste.
6. What chemicals are needed to perform these processes?
7. What kinds of and how much waste do these processes generate?
8. Identify chemicals that are used because of process requirements/specifications.
9. Identify chemicals or processes for which you would like to find alternatives.

Determine How Much Waste is Being Generated

The next step is to determine waste generation qualities, quantities, and sources. This will help focus the assessment on the activities that are causing the largest impact on laboratory operations. Connecting the chemical usage to the waste generation provides the level of detail necessary to refine the scope of the assessment. It allows you to focus on the items that will have the greatest impact within the laboratory. Most organizations have tracking systems for their regulated waste disposal that can be searched for these kinds of data. Other data sources include the staff who manage waste collection and disposal.

Communicating with the researchers about their waste generation and gathering information about chemical usage can be done simultaneously. The following sample questions address waste generation qualities, quantities, and sources:

1. Identify the largest waste streams over a given period of time.
2. Are the waste streams generated from multiple processes or just one?
3. Are the waste streams sanitary, hazardous, or radioactive?
4. Identify problem waste streams, i.e., those that are difficult to dispose of or challenging to manage, or that pose a high risk to staff or the laboratory.
5. Identify waste streams that may be treatable.

Document the Scope of the Assessment

Now that you have evaluated the chemical usage and waste generation, refine the scope of the assessment and document it in the assessment worksheets. First, write a detailed scope statement in Worksheet 1 that includes

- Background information about the group participating in the assessment
- General description of the research activities and waste being generated
- Any additional information on the scope of the assessment
- Areas that are not being included in the assessment and why
- A list of previously accomplished pollution prevention successes.

Next, document the chemical usage and waste generation data gathered in Worksheet 2. Provide these data at the level of detail that is most useful to the assessment. For example, it is not necessary to include areas not being considered within the assessment. If you do not have exact values, use ranges of estimates within an order of magnitude.

See Chapter 7 for specific instructions on each of the worksheets and the appendix for an example of completed worksheets.

Research Similar Pollution Prevention Successes

Now that you have refined the scope of the assessment and know the processes, chemicals, and waste streams on which you want to focus, determine if anyone has previously investigated these same areas. Keep in mind that the technologies for performing "green" research have been improving dramatically over the last few years. A technique, technology, or substitute chemical that was tried unsuccessfully several years ago may have improved and be worth trying again.

To find this information, seek internal resources first, e.g., your pollution prevention program or specialists, environmental compliance or waste management experts, and other researchers performing similar research. Some form of microscale chemistry may have already been tested within your organization and you can learn from past experiences. The pollution prevention program may be able to provide copies of past assessments that include similar research processes or equipment. The environmental compliance and waste management experts can provide their knowledge of techniques or technologies being used in other laboratory spaces. Other researchers are excellent resources for understanding the benefits and challenges of a new technique or technology.

Next, look externally. Search the Internet, look through technical journals and equipment catalogs, and discuss the topic at professional society meetings. These external resources may provide information on ways that other researchers are minimizing waste generation in their laboratory research. Reviewing equipment catalogs for items that are being advertised as "microscale," "environmentally friendly," or "the latest technology" may provide technical solutions to laboratory waste generating processes. At the same time, take all of the information gathered from the Internet, technical journals, and equipment catalogs with a grain of salt. The technique or technology may be marketed as "environmental" when it is only marginally so, and it might not be applicable to the type of research that you perform.

Complete a Walkthrough of the Assessment Area

The walkthrough of the laboratory spaces involved in the assessment is the beginning of the creative portion of the assessment process. The intent of the walkthrough is to have the assessment team physically observe the laboratory space(s), both their own laboratories and those belonging to other researchers, which are part of the assessment scope. Seeing where and how the work is being done helps the assessment team place their recommendations in the context of the researchers who perform the work. The assessment team's walkthrough also helps identify additional issues that the team leader could not uncover through the review of chemical usage and waste generation.

The walkthrough involves the team leader giving a tour of the laboratory spaces included in the scope of the assessment. The team leader coordinates the tour with the researchers in the laboratory spaces so that they expect visitors and understand the purpose of the visit.

The assessment team and the researchers should be aware that this walkthrough is not an audit but more of a self-assessment. It is vital that the assessment team does not go into the walkthrough with an "auditing attitude." The questions from the assessment team should not be judgments on how the work is currently being done. The team must assume that the researchers are working diligently to perform the best research that they can. The walkthrough should help the team understand the research and identify any opportunities for pollution prevention. With the assessment team focused on pollution prevention, it will be able to identify new ideas that might be feasible substitutions for current approaches in addition to opportunities that the researchers wanted to pursue but lacked the resources to do so. The following are examples of simple questions that can be useful in understanding the laboratory work and identifying potential waste reduction opportunities during a walkthrough:

1. What is this piece of equipment used for?
2. What is this chemical used for?
3. Do you know of any substitutions for this chemical that are less hazardous?
4. Does this experiment have to be performed at this scale or could it be smaller?
5. What is in this container, drum, etc.?
6. Why do you do it that way?
7. How do you think you could minimize this waste stream?
8. Can this be reused?

The assessment team needs to listen to the restrictions or reservations in the researchers' responses to these questions. They may have specific requirements to meet for the experiment or may have tried to change the process in the past. Listen for the difference between real limitations and the feeling that "This is the way we have always done this." If pollution prevention opportunities are identified for people who are not currently on the assessment team, it is important to include them in the remainder of the assessment process so that they can provide input and own any suggestions made to improve their process.

The team should also look at the following areas: secondary activities related to the scope, such as maintenance and washing of equipment; ancillary waste streams, such as pipette tips, glassware, batteries, and rags; nonchemical waste streams, including air emissions and wastewater; and use of raw materials, such as water and energy.

Example

Since the processes in the Molecular Biosciences Department were changing at the time of the assessment, the assessment team leader decided not to base their waste generation only on past waste management records. Instead, the team leader interviewed each of the research staff who worked in their individual laboratories to find the answers to three questions:

- What is your production of low-level radioactive waste, low-level radioactive mixed waste, and hazardous waste?
- What three processes contribute the largest volume to the waste streams?
- What is sent through to the process sewer?

In addition to the interviews, the team conducted several walkthroughs of the department's laboratories to identify any previously unnoticed waste streams, such as those that might come from support functions (laboratory cleaning and maintenance, for instance). From this approach, 18 different processes with varying waste streams were identified, ranging from DNA analysis to tissue cultures to chemical storage.

This information was documented on interview forms, as well as summarized on the pollution prevention opportunity assessment worksheets. Copies of these can be found in the appendix.

Brainstorm

Chapter 4

Creating Solutions

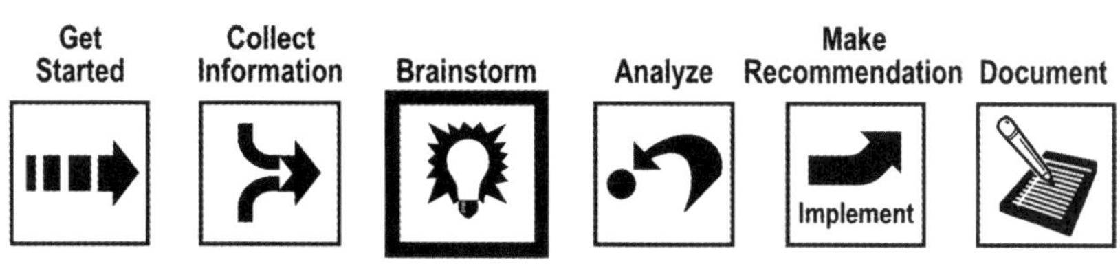

1. *Invite the participants.*
2. *Prepare the participants.*
3. *Conduct a brainstorm session.*
4. *Select opportunities.*
5. *Assign actions.*

Brainstorm

Now that you have collected and documented information about research laboratory operations, you can begin the most creative portion of the assessment: brainstorming opportunities. Brainstorming is a process of creativity followed by action. The essential steps include inviting and preparing the participants, holding the brainstorming session itself, and then assigning actions.

Invite the Participants

At a minimum, the invitation list for the brainstorming session should include the team members. In addition to the team members, invite others who may have useful knowledge of the laboratory activities, such as technicians, waste disposal experts, procurement staff, laboratory monitors, environmental experts, students, and managers. Make sure that those directly involved in the activity you are reviewing will be in attendance, since they have the best knowledge of the activity and of what others in their field may be doing. Also, it is useful to include experts in other fields in order to bring in new viewpoints.

Prepare the Participants

For the brainstorming session to be effective, those involved need to be well informed about the activity. Send out the activity description, chemical usage, and waste generation information that you have collected in advance.

One technique that works well for busy groups is to conduct the walkthrough (as described in Chapter 3) with the core team immediately before the brainstorming session. Devoting a block of time to the assessment, such as two hours for the walkthrough and two hours for the brainstorming and planning session, helps everyone focus on the assessment while the information is still fresh in their minds. A walkthrough or data collection effort a few days or even a week in advance works well, giving ideas time to ferment.

Conduct a Brainstorm Session

The goal of the brainstorming session is to create a list of ideas to reduce material use, reduce waste, and save money. Determining if an idea is practical or possible is done later during the analysis phase. The brainstorming session should be a free flow of ideas and solutions. Bring all the information collected thus far about laboratory activities to the brainstorming meeting, including waste streams, material use, and activity description.

One person should act as the facilitator, recording the ideas. The facilitator guides the meeting to make sure the focus is on source reduction and recycling and reusing materials. The facilitator also makes sure that the group considers all the waste streams involved. The pollution prevention specialist may be a good choice for facilitator if familiar with laboratory operations. Facilitators who are not researchers are more likely to pose basic questions that no one in the laboratory has thought to ask, such as, "Why do you do it that way?" or "Why do you do that at all?" Familiarity with processes can make it difficult to see what a pollution prevention specialist may see as a solution or option.

Keeping the brainstorming session focused on pollution prevention is important to make progress. However, thoughts need to be free, letting the conversation flow to encourage creativity. No idea should be considered silly. The goal of the brainstorming session is to generate as many potential pollution prevention ideas as time allows. Reserve judgment on the feasibility of an idea until all the ideas are verbalized.

As the group discusses each waste stream, the facilitator develops a list of possible opportunities to reduce waste and material usage. General ideas are more useful at this stage because details will be worked out later in the analysis phase.

Select Opportunities

Once the team has generated the ideas and the facilitator has recorded them, the next step is to review and sort the list. Some ideas may be obviously impractical and should be rejected from further study. Other ideas may be so similar that they can be combined into a single item. The focus also can be placed on opportunities that reduce more difficult or hazardous waste streams. Narrow down the list to a reasonable number; e.g., 3 to 10 workable ideas (pollution prevention opportunities). For further investigation, gather more information about each idea and analyze the information.

Prepare a list of the pollution prevention opportunities that were not pursued and note the reasons why they were not considered at this time. This list is included in the documentation on Worksheet 4 and offers useful information regarding potential future pollution prevention opportunities that may be considered during continuous improvement efforts that occur after the assessment. This list also may provide a starting point for future assessments by identifying options that might be worth pursuing and those that were determined impractical or technically infeasible and therefore not worth further investigation.

Assign Actions

Assigning the investigation of each workable pollution prevention opportunity is crucial. The analysis of these opportunities is the most time-consuming and important part of the pollution prevention opportunity assessment. The team leader either assigns these jobs to team members or takes on responsibility for most of the analysis and assigns the information-gathering tasks to others. Before the brainstorming meeting ends, the team leader should have assigned each opportunity to a person who is responsible for its investigation. The team then will leave with actions to accomplish as the next phase of the assessment.

Example

The brainstorming session for the Molecular Biosciences Department's assessment took place several weeks after the walkthrough. The assessment team leader used the time between the walkthrough and the brainstorming session to document and summarize the information from the walkthrough and related interviews into a summary of the Department's research activities, materials, and wastes. The brainstorming session, including the entire assessment team, was facilitated by the pollution prevention specialist and generated about 25 potential opportunities. The group then prioritized the opportunities, grouping some together, rejecting some as not worth further study, and highlighting those that could be valuable. This resulted in nine opportunities considered worthy of further analysis:

- Reduction of wastes from protein assays through microscale techniques
- Outsourcing of custom oligonucleotide synthesis to an external laboratory
- Elementary neutralization of certain corrosive waste from DNA analysis
- Ethanol recovery and reuse through purchase of a solvent distillation system
- Installation of an automatic floor wash-down room
- Procurement of animal bedding containing recycled material
- Recovery or recycle of HPLC solvents
- Establishment of a formal chemical procurement policy to reduce chemical purchases
- Creation of an exit procedure to gather information about chemicals and waste materials from staff before they leave the Molecular Biosciences Department for retirement or other jobs.

The pollution prevention specialist and the assessment team leader assigned actions for each opportunity to members of the team for further investigation.

Analyze

Chapter 5

DEVELOPING IDEAS

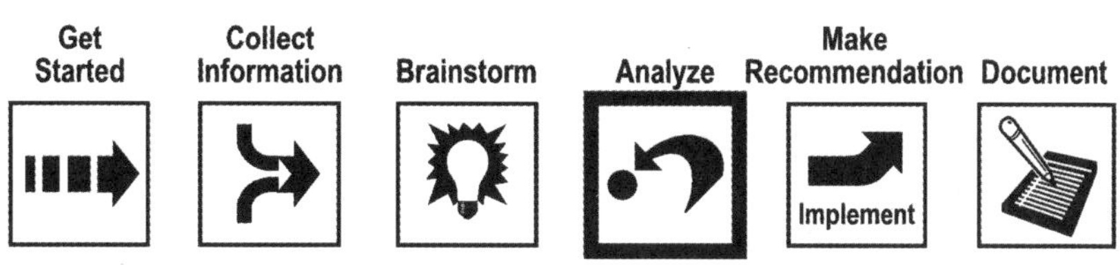

1. *Describe the opportunities.*
2. *Perform a qualitative analysis.*
3. *Perform a quantitative analysis.*
4. *Determine the amount of waste reduced.*
5. *Evaluate cost avoidance.*
6. *Determine implementation cost.*
7. *Calculate payback.*
8. *Calculate return on investment.*

ANALYZE

With brainstorming and prioritization complete, you now have a selection of potential waste-reducing opportunities that may save money for your research organization. To determine if these ideas are worth implementing, you need to conduct an analysis of each opportunity, preferably both qualitative and quantitative analyses.

Describe the Opportunities

The analysis of the pollution prevention opportunities begins with writing out the opportunities in detail. In the worksheets provided in Chapter 7, a description of the current practice is used to define the activity. Next, the pollution prevention opportunity is described as the proposed action to replace the current practice. This includes a qualitative description of the expected benefits and any challenges that may arise. The purpose of documenting the current practice and recommended action is to provide enough detail to solicit implementation funding, if needed, and to explain the opportunities analyzed to those who read the assessment.

Perform a Qualitative Analysis

Some form of qualitative and quantitative analyses will need to be performed for both the current practices and the recommended actions. The following steps can help develop the qualitative analysis:

1. Determine what would be required to implement this opportunity, such as buying new equipment or supplies, conducting research into a new analytical method, or writing a new procedure.
2. If you will need to buy equipment, contact vendors to determine what equipment is needed, its specifications and efficiency, ancillary equipment or software required, implementation requirements, and cost. Talk to several vendors to get a range of options and requirements.
3. Determine how much labor or training will be required to install and use the new equipment.
4. If you are doing a procedure or process change, determine who will be researching the new technique, what materials are needed and their costs, how many hours are needed to develop or research the new method, and training needed to implement the new method.
5. Evaluate and document other benefits besides waste and cost reduction, such as increased safety, reduced worker exposure, improved research capacity, higher quality data collection, marketability of an innovative technique, and any others.

Perform a Quantitative Analysis

The quantitative analysis evaluates the potential waste reduction and cost savings and answers four essential questions:

1. How much waste will the opportunity reduce, either annually or one-time?
2. How much money will the opportunity save, either annually or one-time?
3. What will be the up-front cost to implement the waste reduction measures?
4. What will be the payback or the return on investment?

Determine the Amount of Waste Reduced

There are two kinds of waste reduction: annual and one-time reductions. Annual reductions come from changing an ongoing process so that savings will be yielded for many years after the new technology or technique has been implemented. An example of an annual reduction is changing an analytical process from full-scale to microscale for all future analyses. One-time savings come from making changes to a single action, such as sending excess chemicals from a laboratory cleanup to another laboratory or university for reuse, instead of disposal.

To determine the amount of waste that will be reduced from each opportunity, the following steps should be taken:

1. Review the amount of waste generated now (from a routine process) or estimate the amount of waste to be generated from a one-time activity.
2. Estimate how much waste will be reduced from this pollution prevention opportunity, based on the information gathered when developing the pollution prevention idea. The reduction may be based on efficiency of new equipment, percent reduction in the amount of sample or materials required, or amount of materials recycled or reused.
3. Determine what type of waste will be reduced, as this can greatly impact costs. Waste is usually either hazardous (sent to a qualified hazardous waste disposal company) or nonhazardous (sent to a sanitary landfill or sanitary sewer). Hazardous waste is much more expensive and time-consuming to dispose of than nonhazardous waste. Some laboratories may generate radioactive waste, including radioactive mixed waste (hazardous waste that is also radioactive). These wastes are the most expensive and troublesome for laboratories to dispose. The waste form is also important: solid waste (including liquids that are lab-packed, that is, packaged as a solid waste), liquid waste going to a process or sanitary sewer, and gaseous air emissions.
4. Determine which wastes will be eliminated or reduced, as well as any reductions in the toxicity of a waste (such as reducing hazardous waste to nonhazardous waste).

Evaluate Cost Avoidance

Total cost avoidance comes from three types of savings:

Cost avoidance = waste disposal savings + raw material purchase savings + labor savings

Waste disposal savings—the amount of waste reduced, multiplied by the unit cost of disposing of those wastes. Waste disposal savings should include the cost to collect, log, handle, package, and ship the waste.

Raw material purchase savings—the amount of new materials you would avoid having to purchase because of reuse or reduced use, multiplied by their unit costs.

Labor savings—the amount of labor time reduced, multiplied by a unit labor rate. Labor savings can arise from process efficiencies from new equipment or methods, reduced time spent ordering materials or maintaining equipment, and reduced compliance and safety requirements.

Determine Implementation Cost

Implementation cost is the one-time expense involved in making the change to the process. Implementation cost comes from two types of expenses:

$$\textit{Implementation cost} = \textit{equipment purchase costs} + \textit{labor costs for implementation}$$

Equipment purchase costs—the costs paid to the vendors to buy new equipment, plus any software or ancillary equipment, taxes, purchase adders, or additional direct costs.

Labor costs for implementation—the number of hours required to implement this new activity, multiplied by a labor rate. Labor costs for implementation include costs for research of a new method, changing permits or other regulatory permissions, selecting and installing equipment, training, writing new procedures, and providing any reports required for start-up.

Calculate Payback

The final analysis step is to determine if it is financially worthwhile to implement a waste reduction project.

For projects with *annual savings,* the simplest kind of cost analysis is simple payback:

$$\textit{payback (years)} = \frac{\textit{implementation cost (\$)}}{\textit{annual cost avoidance (\$/year)}}$$

The payback is the number of years or the fraction of the year needed to repay the initial investment in implementing the opportunity. Most businesses consider projects with up to a three-year payback worthwhile to implement. However, some businesses look only at one-year paybacks, while others think more long term, for instance, up to 10-year paybacks.

For projects having *one-time savings,* subtract the implementation cost from the one-time savings:

$$\textit{payback savings (\$)} = \textit{cost avoidance (\$)} - \textit{implementation cost (\$)}$$

If the result is positive (> zero), it is worthwhile to implement.

Calculate Return on Investment

Businesses that use a percent return-on-investment (ROI) measurement often require at least a 33% ROI, though some require as much as a 100% return and others as low as 10%.

To calculate ROI, inverse the payback calculation and multiply by 100 to get a percentage:

$$\textit{\%ROI/year} = \frac{\textit{annual cost savings (\$/year)}}{\textit{implementation cost (\$)}} \times 100$$

If you want to take into account the time-cost of money or depreciation, the calculation becomes a little more complex. Here is one way to calculate it:

$$\%ROI/year = \frac{annual\ cost\ savings\ (\$/year) - \dfrac{implementation\ cost\ (\$)}{life\ of\ project\ (years)}}{implementation\ cost\ (\$)} \times 100$$

The *life of project* is how long the new equipment or new process will be used, usually 5 to 20 years.

Example

A detailed analysis was conducted of nine opportunities developed during the brainstorming session for the Molecular Biosciences Department assessment. The analysis of the first opportunity was completed in the following manner:

Opportunity: Reduction of wastes from protein assays through microscale techniques

Current Practice: At least three different groups within the Molecular Biosciences Department routinely measure protein levels in biological samples using either the method of Lowry or Bradford. Current sample throughput is 2000 Lowry's and 1000 Bradford's per year. The Lowry assay generates 4.0 mL hazardous waste/assay and requires ~2 minutes of technical time per sample for set-up and analysis; the Bradford assay generates 5.0 mL waste per sample and also requires ~2 minutes of technical time per sample.

Proposed Action: Purchase an automated "microplate reader" which would allow the performance of both assays on a microscale level, reducing the amount of hazardous waste for both to 0.2 mL and reducing handling time to 1 minute per sample.

Calculation of Waste Reduction: The current waste stream from this activity is approximately 13 kg hazardous waste per year. After implementation, this would be reduced to approximately 0.6 kg/year (12.4 kg reduction of hazardous waste).

Calculation of Annual Cost Savings:

Current		
	Labor (101 hr x $85/hr)	$8,585
	Disposal (13 kg x $50/kg)	650
	Reagents	573
	Total	$9,808
Proposed		
	Labor (51 hr x $85/hr)	$4,335
	Disposal (0.6 kg x $50/kg)	30
	Reagents	136
	Total	$4,501
Annual Cost Savings:		$9,808 - $4,501 = $5,307

Calculation of Implementation Cost and Payback:

Implementation Cost		
	Microplate Reader	$15,000
	Labor and Training (8 hr x $85/hr)	680
	Total	$15,680
Payback (Implementation Costs/Annual Cost Savings):		$15,680/$5,307 = 2.9 years

Make Recommendations and Implement

Chapter 6

Deciding What to Implement

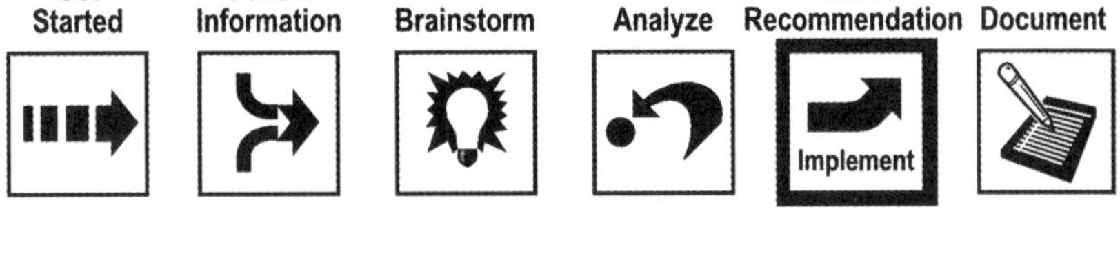

1. *Select opportunities for implementation.*
2. *Document the implementation strategy.*
3. *Present results to management.*
4. *Implement recommended activities.*

MAKE RECOMMENDATIONS AND IMPLEMENT

Now that the team has identified and researched the pollution prevention opportunities relevant to the research work, the next action is to decide which of these should be implemented. This step leads to the most important activity of the entire assessment process: implementation. Pollution prevention opportunity assessments provide the technical and cost savings bases for recommending changes to laboratory techniques or equipment. This section discusses how to consolidate the information that has been collected and use it to garner implementation funds. Implementation of the pollution prevention opportunities identified in the assessment is a crucial step in the culture change that begins once researchers find that preventing pollution provides opportunities to advance their research capabilities.

Select Opportunities for Implementation

To select opportunities to implement, first review the payback calculated for each. Consider first those opportunities that have little or no cost associated with them. Next, consider any opportunity that has a payback of three years or less as usually worthy of implementation. Items with a payback period longer than three years may be worth implementing as well, depending on other benefits. Reasons to implement opportunities regardless of payback include: impact on a high-profile research project, reduction of a high volume of waste, reduction of a highly dangerous waste, reduction of risk and liability, reduced worker exposure, simplicity of implementation, high interest by researcher, or broad applicability in other laboratories.

Document the Implementation Strategy

Now that you have researched the pollution prevention opportunities and selected the ones you want to implement, it is time to document these recommendations. First, summarize the numerical results of the opportunities by listing the opportunity title, waste type reduced, annual waste reduction, estimated annual cost savings, estimated implementation cost, and payback.

Next, summarize the opportunities and document the implementation recommendations identified by the assessment team. This summary should include the identification of opportunities that are not being pursued further and a prioritized list of the opportunities that are considered worthy of implementation. The opportunities that are being considered for implementation need to be organized as follows:

- Opportunities that are being pursued immediately because they are free
- Opportunities that are being pursued immediately and the source of the implementation funds
- Strategy for identifying implementation funds and other resources for implementation of the remaining opportunities on the list.

See Chapter 7 for specific instructions on each of the worksheets and the appendix for an example of completed worksheets.

Present Results to Management

One of the keys to the success of a pollution prevention program is the involvement of managers or organizational leaders. Sharing the results of a pollution prevention opportunity assessment is one way to share the research organization's progress in improving the bottom line as well as demonstrating that researchers are proactive in finding solutions to their environmental challenges.

The recommendations of the assessment should be presented to key management. At this time, provide managers with a list of projects that are going to be implemented because they are easy and inexpensive, show them the effect that these projects have on the bottom line, and provide a list of items that are worthy of being implemented but need implementation funds. Emphasize any improvements to research, risk reduction, and reduced compliance costs. Demonstrating that you have researched and prioritized the opportunities based on their financial benefit to the company typically provides managers the data they need to justify investment. Additionally, proactively identifying and researching pollution prevention opportunities and showing that you have implemented all of the projects within your budget demonstrates commitment to saving the organization money, improving safety, reducing risk, and investing in the environment.

Implement Recommended Activities

As mentioned above, the first opportunities that should be implemented are those which are free or nearly free. The next step is to identify sources of funding for the opportunities that are determined worthy of implementation but do not have the resources to implement. First, consider funding these opportunities on research project funds, using the payback information as a selling point to the funding clients of those research projects. The next route is to talk with the management of your organization to see if there are internal funds that would support pollution prevention investments. Next, contact your pollution prevention specialists to find out about any implementation programs that might be available. Finally, if internal or client resources are not available, look externally on the Internet for government and other resources that might support pollution prevention implementation projects.

Example

Nine pollution prevention opportunities were identified as part of the Molecular Biosciences Department assessment. Of these, four were found to yield quantifiable reductions in waste streams. These four opportunities were technical in nature and had technical solutions. They were recommended for implementation because they (1) had good paybacks, (2) were simple to implement, or (3) were beneficial for the Department because they reduced waste. Two other recommended opportunities involved administrative procedural changes that were difficult to quantify, but represented good management practices that would result in less risk and liability. These also were recommended for implementation. Finally, three opportunities were not recommended for implementation because they resulted in very small waste reductions, the recommended items were not available, or the current process was already being replaced.

A proposal for implementation funds was completed for one of the opportunities, and the remaining recommended opportunities were pursued through management or through permission from waste management organization.

Document

Chapter 7

Recording Your Success

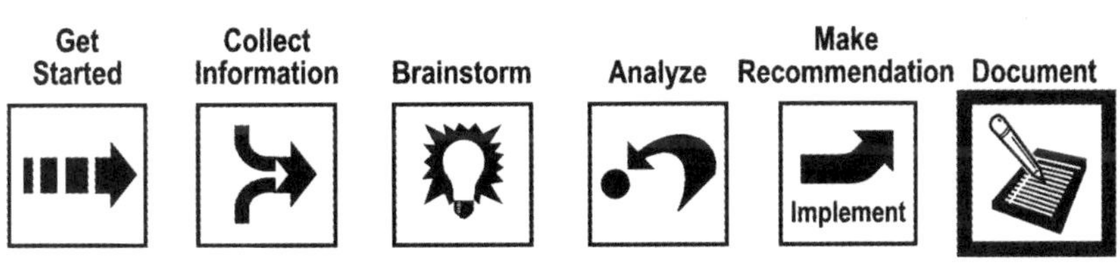

1. *Complete pollution prevention opportunity worksheets.*
2. *Provide completed report to management.*

DOCUMENT

The task often considered the least inspiring—documentation—is also one of the most important. It is important that the assessment be documented *as it is being performed.* The final documentation is the completed worksheets, which become the record of the pollution prevention opportunity assessment. Properly completed, the worksheets summarize the recommended actions, the next steps, and the point-of-contact information in enough detail that the ideas can be implemented. Additionally, the documentation of one assessment can be used as a beginning for the next assessment in a similar research area. Occasionally, assessments will identify opportunities that apply not only to a particular laboratory but also to other laboratories. Other benefits of good documentation include use of documentation to easily prepare proposals for funding and use of the documentation as evidence of a pollution prevention program for regulators and others.

Complete Pollution Prevention Opportunity Worksheets

The following five worksheets have been found to be useful in documenting the pollution prevention opportunity assessment efforts:

Worksheet 1: Team and Activity Description
Used to document the background and scope of the assessment.

Worksheet 2: Activity Flow Diagram
Used to document the materials flow and mass balance within the scope of the assessment.

Worksheet 3: Pollution Prevention Opportunity
Used to document each of the pollution prevention ideas that are being considered for the assessment. There will be one of these worksheets for each idea being analyzed; thus, each assessment will have several Worksheet 3s when the assessment is completed.

Worksheet 4: Pollution Prevention Opportunities Summary
Used to summarize the benefits of the pollution prevention opportunities and to list opportunities that were brainstormed but not included in the assessment.

Worksheet 5: Final Summary
Used to summarize the recommendations and strategies for implementation of the pollution prevention opportunities identified in the assessment.

Each worksheet is discussed in detail in this chapter. To find electronic copies of these worksheets, visit the PNNL Pollution Prevention Web site at http://p2.pnl.gov:2080/p2/p2oa.htm.

Experience has shown that the details provided in a well-documented pollution prevention opportunity assessment can be used in a variety of ways:

- Drafting implementation funding requests
- Providing a methodology (logical steps) for performing future assessments from beginning to end
- Recognizing the research and development laboratory's success in completing the assessment through publications and awards

- Advertising the pollution prevention opportunities that might be applicable elsewhere
- Demonstrating the benefits (such as waste reduction, researcher time saved, and cost savings) associated with performing assessments
- Encouraging other research and development laboratories to perform assessments.

In addition to the worksheets, some assessment teams have included executive summaries to summarize the assessment for management, as well as appendices with detailed information supporting the assessment calculations or process.

Provide Completed Report to Management

Once the worksheets have been completed, it is important to share the assessment results with management and your pollution prevention specialist. Management needs to be made aware of the completed assessment so they know that you are working toward continuous improvement in your research, avoiding environmental compliance issues by acting proactively, reducing waste at the source, and reducing costs. Additionally, management and the pollution prevention specialist may have access to funds to be used for implementation of the pollution prevention opportunities identified in the assessment. The completed report will allow them to help you find funding for new equipment or for time to change processes.

Pollution Prevention Opportunity Assessment
Worksheet 1
Team and Activity Description

Date (1) **P2OA ID Code** **Facility**
Activity

Team Members (*Leader) (2) Telephone Title

Description of Activity to be Examined in this P2OA (3)

Worksheet 1

The intent of Worksheet 1 is to provide the background and scope of the assessment. See Chapter 2 for direction on how to collect the information needed for Worksheet 1. To complete Worksheet 1:

(1)

- Title the assessment (typically named after the research group's name, their building, and/or their type of research).

(2)

- Provide the team members' names and contact information.
- Flag the team leader as primary point of contact.
- Provide brief titles for team members, such as pollution prevention specialist, researcher, etc.

(3)

- Provide background information about the entity being assessed (e.g., type of work, number of people, name of the group, number of current projects, etc.).
- Provide general description of waste generated or research activities that will be evaluated.
- Provide scope of assessment (if different than above).
- Provide a list of the areas *not* being considered in this assessment and why (can be general, such as, "no waste streams under 1 kg were considered," or specific, such as, "the Biology department staff also housed in the Chemistry building were not included in the scope of this assessment").
- Provide a list of previously accomplished pollution prevention activities.
- If more detail is available, this is a good place to put one-paragraph descriptions of the laboratories being assessed and all the processes that are included within the scope.

Pollution Prevention Opportunity Assessment
Worksheet 2
Activity Flow Diagram

Date P2OA ID Code Facility

Activity

Chemical and Radioactive Inputs

Name	Qty.
1	

Material Inputs

Name	Qty.

Energy Inputs

Name	Qty.

Activity
Activity Time Period

2

Product or Result Output

Name	Qty.
3	

Hazardous Waste Output

Name	Qty.

Nonhazardous Waste Output

Name	Qty.

Radioactive Waste Output

Name	Qty.

Mixed Waste Output

Name	Qty.

Other

Name	Qty.

Worksheet 2

The intent of Worksheet 2 is to demonstrate a materials flow and mass balance within the scope being assessed. A mass balance is not always feasible or relevant for research and development pollution prevention opportunity assessments, so it is optional for this worksheet. The input materials and the waste output, including quantities, are the important portions of this worksheet. See Chapter 3 for direction on how to collect the information needed for Worksheet 2. To complete Worksheet 2:

1

- List and provide quantities of key chemical, water, energy, materials, and other relevant resources being addressed within the scope of the assessment.

2

- Provide a short description of the research being performed.

3

- List and provide quantities of key waste streams being addressed within the scope of this assessment.

Pollution Prevention Opportunity Assessment
Worksheet 3
Pollution Prevention Opportunity

Date **P2OA ID Code** **Facility**
Activity (1)
P2O No. 1 **P2O Title**

Current Practice (2)

Proposed Action (3)

Calculation of Waste Reduction and/or Energy Savings (4)

Calculation of Annual Cost Savings (5)

Calculation of Implementation Cost and Payback (6)

Vendor/Contact Information (7)

Worksheet 3

The intent of Worksheet 3 is to document the pollution prevention opportunities that are being considered for the assessment. There should be one Worksheet 3 for every pollution prevention opportunity that is analyzed. Typically, the number of Worksheet 3s range from 3 to 10, depending on the scope of the assessment, complexity of the pollution prevention opportunities suggested, and how much pollution prevention was performed previously. See Chapters 4 and 5 for direction on how to collect and calculate the information needed for Worksheet 3. To complete Worksheet 3:

(1)
- Uniquely title each pollution prevention opportunity

(2)
- Provide a description of how the current research process and/or activity is being performed, how much waste it is generating, what type of equipment it is using, the amount of researcher time required, the types of chemicals used, and what the work is trying to accomplish.

(3)
- Describe the proposed action including the details on the changes that would need to be made; what type of equipment would be required (if any); whether it would reduce waste, time, money, or hazard level; what activities would need to be performed for implementation to take place, etc. Provide enough detail here that this could be reformatted into an implementation funding proposal if necessary.
- Describe the activities that will need to be performed for implementation and then list each item and its cost.

(4)
- Describe the waste type being minimized or eliminated and provide the quantity; show any calculations that are relevant.

(5)
- Provide detail on the annual costs of the current practice and proposed action (not including implementation costs). Calculate annual cost savings associated with changing to the proposed action. Note that the cost includes the cost of managing and disposing of waste, changes in staff time to perform the work, changes in cost of purchased items, and any other relevant costs that can be identified (see Chapter 5).

(6)
- Calculate implementation cost (see Chapter 5).
- Calculate payback (see Chapter 5).

(7)
- Provide point of contact for this particular opportunity (team leader or otherwise) and any internal or external resources that may need to be contacted if this opportunity were to be implemented.

Pollution Prevention Opportunity Assessment
Worksheet 4
Pollution Prevention Opportunities Summary

Date **P2OA ID Code** **Facility**
Activity

P2O No.	P2O Title	Waste Class Reduced	Annual Waste Reduction or Energy Savings	Estimated Annual Savings	Estimated Implementation Cost	Payback
1	(1)	(2)	0	$	$	years
2			0	$	$	years
3			0	$	$	years

Other Brainstorming Opportunities

(3)

Worksheet 4

The intent of Worksheet 4 is to summarize the benefits identified in Worksheet 3. This table has no new data in it, just the titles of opportunities, waste types reduced, waste reduction, cost savings, implementation cost, and payback information calculated on Worksheet 3. The only new information on this worksheet is a list of brainstorming opportunities that were not pursued and an explanation of why they were not considered at this time. To complete Worksheet 4:

1

- Copy the title of each opportunity into one row of the table.

2

- Copy the waste type reduced, waste reduction quantity, cost savings, implementation cost, and payback final calculations into the appropriate columns.

3

- List any opportunities that were discussed in your brainstorming session but were not analyzed. Provide a reason why they were not pursued at this time (this helps document any ideas you ruled out for others reading your assessment in the future).

Pollution Prevention Opportunity Assessment **Worksheet 5** **Final Summary**		
Date	**P2OA ID Code**	**Facility**
Activity		

Proposed Opportunities and Discussion ①

Recommendations and Schedule for Implementation ②

Worksheet 5

The intent of Worksheet 5 is to provide recommendations for implementation and a strategy for implementing and/or identifying resources needed to implement. See Chapter 6 for direction on how to collect the information needed for Worksheet 5. To complete Worksheet 5:

(1)

- Briefly summarize the pollution prevention opportunities (on Worksheet 3) that were evaluated.

(2)

- Summarize the opportunities that are recommended for implementation and why. Describe the steps that will be taken to implement each recommended opportunity. Mention any opportunities that were recommended and already implemented during the assessment.
- Briefly summarize the opportunities that are recommended for implementation but need funding/resources from a source external to the group. Describe the recommended path forward for pursuing implementation funds. Identify point of contact for the implementation of this opportunity.
- Briefly summarize the opportunities that are marginally recommended for implementation, explain why they are only marginal, and describe the best path forward as it is seen now. Also, provide a date to re-evaluate these opportunities to see if their chances of implementation improve over time.
- Optional—provide 'lessons learned' on the assessment that can be used as recommendations to the pollution prevention specialist on the next R&D assessment.

Example

The Molecular Biosciences Department completed the worksheets, but took the documentation one step further. The team leader attached appendices to the assessment with detailed information about the data collection and the opportunity analysis. An executive summary and formal letter summarizing the assessment results were prepared and submitted to management. This process assisted in the implementation of the opportunities after the assessment was completed. The complete document, including worksheets and executive summary, is included in the appendix.

Additionally, the Molecular Biosciences Department used the report successfully in a subsequent compliance audit. With the report and the actions it had taken, the department was able to demonstrate that it had a formal, analytical approach to reviewing and reducing its waste streams and that it acted on opportunities to prevent pollution. The regulators were impressed with the new proactive approach from the Molecular Biosciences Department, and there were no regulatory findings.

The PNNL pollution prevention team also documented the results from the Molecular Biosciences Department with several articles in company newsletters, submission of the assessment for an award, and presentation of the results in a paper at a pollution prevention conference. These gained recognition for the Molecular Biosciences Department staff and for PNNL as a proactive organization.

Conclusion

Chapter 8

ACHIEVING A POLLUTION PREVENTION CULTURE

CONCLUSION

Completing a pollution prevention opportunity assessment is both a first step towards preventing waste and an essential part of a complete pollution prevention program. An assessment can be conducted at a grassroots level, with researchers and pollution prevention specialists diving into their work processes, developing creative techniques, and finding new technologies to reduce their waste streams. However, by combining assessments with a complete pollution prevention program, you can build a strong environmental management system and the beginnings of an environmental culture change within a research organization.

Ultimately, conducting formal pollution prevention opportunity assessments gives structure to a creative process and formalizes an environmental consciousness that most research scientists and staff have by nature and by training. It provides the time, resources, and framework for researchers to create ideas, conduct analyses, and make recommendations to reduce waste. By implementing these recommendations, they can improve their processes in a way that is useful to clients, managers, regulators, and themselves. Assessments make the process of reducing chemical and material use and taking a step towards long-term reduced environmental impact both achievable and desirable.

Example

In the year since the assessment was completed, the Molecular Biosciences Department attained and retained a culture change. In addition to implementing the opportunities discovered in the formal assessment process, it has begun to continually look for pollution prevention opportunities in daily work. The department staff and management actively and routinely seek funding for new technologies that reduce waste and improve the quality of their research.

Since the assessment was initiated, the Molecular Biosciences Department has purchased several new imagers that have reduced its dependence on radioactive tracers and photographic processes that generate radioactive and hazardous waste. It has purchased several types of microscale equipment to reduce the volumes of samples and waste at the source. It has increased its recycling programs of various materials. The Molecular Biosciences Department staff, from student interns to management, have developed, championed, and implemented pollution prevention ideas of all kinds. The pollution prevention culture they developed has inspired other researchers in the building, who have also begun to seek out new techniques and technologies that reduce waste.

Performing the pollution prevention opportunity assessment led the Molecular Biosciences Department to a new way of thinking. The department staff now look at existing and new activities more critically to determine ways to do their research that generate less waste. They are now equipped with the tools to analyze new ideas for cost avoidance potential and other, less tangible benefits, including reduced risk, improved compliance, and worker safety. Pollution prevention has helped make the Molecular Biosciences Department a more competitive and viable research organization.

Definitions & Acronyms

Implementation cost: the one-time expense involved in making the change to your process

Molecular Biosciences Department: a research and development organization at the Pacific Northwest National Laboratory

Payback: the length of time for cost avoidance to equal the investment cost of a project

PNNL: Pacific Northwest National Laboratory

Pollution prevention: reduction of waste through source reduction and recycling/reuse

Pollution prevention opportunity assessment: a systematic, documented approach to determining where pollution prevention opportunities exist and the best methods to implement those opportunities

R&D: research and development

Return on investment (ROI): the percent of the annual savings of the total investment cost

Reuse/recycle: reusing chemicals and materials, or reprocessing or recycling instead of disposing of them

Source reduction: reducing waste at the source before it is generated

Treatment: reducing the toxicity or volume of waste before safely disposing of it

Example of Final Assessment Report

Appendix

Pollution Prevention Opportunity Assessment

PNNL-96-05

Molecular Biosciences Department
Health Division
Battelle - Pacific Northwest National Laboratory

August 14, 1996

Table of Contents

Executive Summary

This Pollution Prevention Opportunity Assessment (P2OA) for the Molecular Biosciences Department (MBD) was undertaken as part of the Health Division's FY96 ES&H Plan. The Team undertaking the task consisted of a professional staff member from MBD (KM Groch), representatives from Environmental Compliance (JA Engel-Cox, KM Fowler, and RR Nielson), Waste Management (M McCulloch), the MBD's Administrator (RT Thomas), and the Health Division's ES&H Officer (JY Young). Financial support for the team's activities were supplied through DOE-RL EM Pollution Prevention Funding. The team was formed in April of this year, held its first meeting on April 23, and finished its work on August 14. The team followed the format described in "Pollution Prevention Opportunity Assessments: Guidance for the Hanford Site" (JA Engel, Westinghouse Hanford Pollution Prevention Group, Richland, WA; WHC-MR-0472. Rev.0) with several modifications described in the report.

The scope of the P2OA included 1) all MBD research activities in the 331 Building and Annex, 2) research-related support activities in the facilities under the direct administrative control of MBD and the Health Division, and 3) MBD business practices impacting on the generation and management of waste streams. The results of the efforts can be summarized as follows:

- A total of nine Pollution Prevention Opportunities were identified for further analysis. Of these, the team recommends implementation of six. If fully implemented, the team estimates that these opportunities would result in an annual waste reduction of ~87 kg (~1/2 the department's current waste stream) and an annual cost savings of ~\$33,000.

- During the course of its activities, the team found external funding for implementation of one of the opportunities (P2O No. 1, "Reduction of Wastes from Protein Assays). Capital funds (~\$9,500) were obtained for the purchase of an automated "Microplate Reader" which will reduce the waste stream from protein assays by ~95% and result in an annual cost savings of ~\$5,300.

- A relevant finding not directly related to the performance of this P2OA were the very significant efforts in waste reduction undertaken by the MBD in FY95. Most importantly, a substantial capital investment was made in a Molecular Dynamics FluorImager along with extensive efforts by MBD staff to modify research protocols to fully utilize this device. As a consequence, the use of ^{32}P was essentially discontinued with a substantial reduction in radioactive wastes as well as the costs associated with the use of radioisotopes.

In conclusion, the team would like to acknowledge and thank all of the staff of the MBD who freely gave their time to the effort. Moreover, the team feels it is important to commend MBD management for its support and encouragement.

POLLUTION PREVENTION OPPORTUNITY ASSESSMENT
WORKSHEET 1
TEAM AND ACTIVITY DESCRIPTION

Date: August 14, 1996 **P2OA Code:** PNNL-96-05 **Facility:** 331 Building & Annex

Activity: Biological Research within the Molecular Biosciences Department (MBD)

Team Members (*Leader)	Telephone	MSIN
Kevin M. Groch*	376-3043	P7-53
Jill A. Engel-Cox	372-0307	P7-79
Kimberly M. Fowler	372-4233	K8-03
Margaret McCulloch	376-9840	P7-57
Robert R. Nielson	375-4536	K9-44
Ronald M. Thomas	376-3999	P7-58
John Y. Young	375-2600	K4-16

Description of Activity to be Examined in This P2OA

The Molecular Biosciences Department (MBD) consists of 17 professional staff members involved in a variety of basic biological research activities. At this time, 11 of these staff members are actively performing laboratory work in the 331 Building and Annex. Without exception, each investigator's work is best characterized as the integration of multiple, small-scale processes utilizing a number of reagents which are performed intermittently. Typically, the individual processes result in the production of waste in the range of a few hundreds of grams to several kilograms per year. In many cases, the wastes generated over time by a number of small processes are collected in a single vessel for storage in a Satellite Accumulation Area (SAA). All of the research activities are to be examined in this P2OA.

The performance of a P2OA typically starts with identifying Priority Streams and Activities based on a variety of historical data bases such as procurement records or disposal requests. This was considered unrealistic for two reasons. First, the very nature of the activities in the department make a search of these types of records very labor intensive requiring far more labor hours than available. Second, the MBD is currently undergoing a major reprogramming in terms of research activities which has resulted in substantial changes of professional staff. Moreover, the remaining staff tend to be pursuing lines of investigation which differ from their previous efforts.

As an alternative to the more typical approach, each staff member was interviewed using a standard form to identify 1) their production of LLW, LLMW, and hazardous wastes; 2) the three processes contributing the largest volume to their individual waste streams; and 3) their use of the process sewer system. Due to time limitations, no effort was made to systematically characterize the production of sanitary waste. In addition to the interrogatories, several departmental walk-throughs were performed by the Team as well as MBD management in an effort to identify any unappreciated waste streams from support functions which are the direct responsibility of the MBD. These are summarized as "Miscellaneous Waste Issues Identified by the Team."

POLLUTION PREVENTION OPPORTUNITY ASSESSMENT
WORKSHEET 2
ACTIVITY FLOW DIAGRAM

Date: August 14, 1996 **P2OA Code:** PNNL-96-05 **Facility:** 331 Building & Annex

Activity: Multiple processes within MBD

Chemical and Radioactive Inputs

Name	Quantity
ethanol	
xylene	
methanol	
formalin	
sodium hydroxide	
misc. other reagents	
samplcs	

Material Inputs

Name	Quantity
water and detergents	
disposable plasticware	
misc. laboratory materials	
misc. general materials	

Energy Inputs

Name	Quantity

Activity
Biological Research
Activity Time Period
1 year

Product or Result Output

Name	Quantity
Biological Research	

Hazardous Waste Output

Name	Quantity
Annex solids, carcinogens	20 kg
Room 129 processed samples	5 L
Room 112 ethanol	10 L
Room 355 ethanol, reagents	<1 L
Rooms 108-110 ethanol, methanol, formalin	<1 L
Room 361 ethanol, methanol	<1 L
Room 359 ethanol, formalin, xylene, misc. reagents	10-15 L

Nonhazardous Waste Output

Name	Quantity

3

POLLUTION PREVENTION OPPORTUNITY ASSESSMENT
WORKSHEET 2
ACTIVITY FLOW DIAGRAM
- continued -

Product or Result Output -continued -

Name	Quantity

Hazardous Waste Output - continued -

Name	Quantity
Room 312	10-15 L
misc. reagent	
Room 316	10 L
reagents w/ silver	
Room 350	2 L
ethanol, misc. reagent	
Room 304	12 L
processed samples	
buffer w/ methanol	8 L
0.4 *M* NaOH	8 L
processed samples	10-15 L
Room 125	25 L
buffer w/ methanol	
processed samples	10-15 L

Nonhazardous Waste Output - continued -

Name	Quantity

Radioactive Waste Output

Name	Quantity
Room 129	40 L
liquid to RLWS	
rinse from cage wash	? L

Mixed Waste Output

Name	Quantity

Other

Name	Quantity

Total Input Mass = Total Output Mass? NA **kg In =** NA **kg Out**

4

WASTE STREAM INFORMATION

PRIORITY STREAMS AND ACTIVITIES

Annex

Process #1: Whole Animal Carcinogenesis Studies
Waste Type: H, M Estimated Waste Stream Volume: 20 kg/year

This is a fairly complicated process which includes treating experimental animals with carcinogen, following the animals over time for the appearance of neoplasia, sacrificing the animals at an appropriate time, and collecting tissue samples. The waste stream consists of excess carcinogen and solid wastes such as gloves used during carcinogen treatment. Currently, all of the hazardous liquid wastes which constitute a small fraction of the 20 kg/year waste stream are returned to the sponsor for disposal. The remainder of the solid wastes are disposed of as medical wastes through a contract with ENSCO. The group currently does not have an active SAA.

Room 129

Process #1: Whole Animal Metabolism Studies
Waste Type: LLR Estimated Waste Stream Volume: 40 L/year

Description:
This is a fairly complicated process involving the injection of $_{14}$C-labeled compounds into experimental rats, collecting body fluids and tissue samples over time, processing these fluids and samples, and analyzing the samples using a combination of HPLC-based methods and LSC. The waste stream consists of the processed samples including the scintillant (Optifluor), effluents from the HPLC, and any excess liquids produced during the processing. The waste stream also consists of the mild, water-based detergent solutions and water rinses used to clean the animal cages as well as the solid waste from all these activities (*e.g.*, scintillation vials, paper towels, *etc.*).

Comments:
The liquid waste stream is currently being disposed of using the LRWS.

Process #2: Inorganic Phosphate Determination
Waste Type: H Estimated Waste Stream Volume: 5 L/year

Description:
The process is a spectrophotometric analysis of total inorganic phosphorous in biological samples using the Fiske-Subbarow method. The process includes isolating the tissue, processing the tissue using biophysical and biochemical methods, and analyzing the samples. The waste stream consists of the processed samples.

Comments:

WASTE STREAM INFORMATION

PRIORITY STREAMS AND ACTIVITIES

Room 112

Process #1: Northern Blotting
Waste Type: H Estimated Waste Stream Volume: 10 L/year

Description:
The process involves resolving RNA by electrophoretic mobility in an agarose gel and subsquent processing of the gels to electrophoretically transfer the DNA onto nylon sheets. The hazardous waste stream consists of ethanol used to process the nylon for reanalysis.

Comments:

Room 355

Process #1: DNA Isolation and Purification:
Waste Type: H Estimated Waste Stream Volume: <1 L/year

Description:
The is a multistep process consisting of the culture of bacteria containing a gene sequence of in-terest, isolating the DNA by a variety of biophysical and biochemical methods, and the subsequent purification of the crude DNA isolate by a variety of additional biochemical methods The waste stream consists of small volumes (1-2 mL) of a variety of reagents. However, a waste common to most of the individual sub-processes is ethanol.

Comments:
The majority of the waste produced is ethanol used to precipitate the DNA after each individual sub-process.

Room 108-110

Process #1: Fixation of Cells
Waste Type: H Estimated Waste Stream Volume: <1 L/year

Description:
The process involves treating tissue culture cells with a variety of reagents for subsequent microscopic or flow cytometric analysis. The waste stream consists of the ethanol, methanol, or formalin.

Comments:

WASTE STREAM INFORMATION

PRIORITY STREAMS AND ACTIVITIES

Room 365

None (No active SAA)

Room 361

Process #1: Fixation of Cells
Waste Type: H Estimated Waste Stream Volume: <1 L/year

Description:
The process involves treating tissue culture cells with a variety of reagents for subsequent microscopic or flow cytometric analysis. The waste stream consists of the ethanol and methanol.

Comments:

Room 359

Process #1: Histopathological Examination of Tissues
Waste Type: H Estimated Waste Stream Volume: 10-15 L/year

Description:
This is a fairly complicated process involving the collection of tissues from animals, fixing and processing the tissue for analysis, and staining the processed tissues using a variety of reagents. The waste stream consists primarily of 10% neutral buffered formalin, ethanol, xylenes, and stains.

Comments:

WASTE STREAM INFORMATION

PRIORITY STREAMS AND ACTIVITIES

Rooms 312, 316, 350, 354, & 356

Process #1: Custom Synthesis of Oligonucleotides
Waste Type: HEstimated Waste Stream Volume: 10-15 L/year

Description:
The process involves the custom synthesis of short, specific DNA sequences for use in a variety of molecular biological procedures. The process utilizes an Applied Biosystems Model 392 DNA/RNA Synthesizer and prepackaged reagents. The waste stream consists of a mixture of excess reagents as an effluent from the synthesizer.

Comments:
The process runs as a "system" based on the manufacturer's prepackaged reagents.

Process #2: Staining of Polyacrylamide Gels for Heme Proteins
Waste Type: H Estimated Waste Stream Volume: 10 L/year

Description:
The process involves resolving proteins by electrophoretic mobility in a polyacrylamide gel and subsequent processing of the gel to specifically stain heme proteins. The waste stream consists of the staining reagent which contains silver.

Comments:
This is a bench-scale activity used on an as-needed basis.

Process #3: DNA Isolation and Purification:
Waste Type: H Estimated Waste Stream Volume: 2 L/year

Description:
The is a multistep process consisting of the culture of bacteria containing a gene sequence of interest, isolating the DNA by a variety of biophysical and biochemical methods, and the subsequent purification of the crude DNA isolate by a variety of additional biochemical methods The waste stream consists of small volumes (1-2 mL) of a variety of reagents. However, a waste common to most of the individual subprocesses is ethanol.

Comments:
The majority of the waste produced is ethanol used to precipitate the DNA after each individual subprocess.

WASTE STREAM INFORMATION
PRIORITY STREAMS AND ACTIVITIES

Room 304

Process #1: Histone Preparation
Waste Type: H Estimated Waste Stream Volume: 12 L/year

This is a fairly complicated process involving the isolation and analysis of histone proteins. The process includes isolating the proteins and analyzing the samples using HPLC. The waste stream consists of the HPLC samples and effluent.

Comments:

Process #2: Western Blotting
Waste Type: H Estimated Waste Stream Volume: 8 L/year

Description:
The process involves resolving proteins by electrophoretic mobility in a polyacrylamide gel and subsequent processing of the gels to electrophoretically transfer the proteins onto nylon sheets. The hazardous waste stream consists of the transfer buffer which contains 20% methanol.

Comments:
It is possible to switch to ethanol.

Process #3: Southern Blotting
Waste Type: H Estimated Waste Stream Volume: 8 L/year

Description:
The process involves resolving DNA by electrophoretic mobility in an agarose gel and subsequent processing of the gels to electrophoretically transfer the DNA onto nylon sheets. The hazardous waste stream consists of 0.4 N sodium hydroxide used to depurinate the DNA prior to transfer.

Comments:

Process #4: Protein Determination
Waste Type: H Estimated Waste Stream Volume: 10-15 L/year

Description:
The process is a spectrophotometric analysis of total protein in biological samples using the Lowry or Bradford methods. The process includes isolating the tissue, processing the tissue using biophysical and biochemical methods, and analyzing the samples. The waste stream consists of the processed samples.

Comments:

WASTE STREAM INFORMATION

PRIORITY STREAMS AND ACTIVITIES

Room 125

Process #1: Western Blotting
Waste Type: H Estimated Waste Stream Volume: 25 L/year

Description:
The process involves resolving proteins by electrophoretic mobility in a polyacrylamide gel and subsequent processing of the gels to electrophoretically transfer the proteins onto nylon sheets. The hazardous waste stream consists of the transfer buffer which contains 20% methanol.

Comments:
It is possible to switch to ethanol.

Process #2: Protein Determination
Waste Type: H Estimated Waste Stream Volume: 10-15 L/year

Description:
The process is a spectrophotometric analysis of total protein in biological samples using the Lowry or Bradford methods. The process includes isolating the tissue, processing the tissue using biophysical and biochemical methods, and analyzing the samples. The waste stream consists of the processed samples.

Comments:

MISCELLANEOUS WASTE ISSUES IDENTIFIED BY THE TEAM

Animal Resource Center

The 331 Building contains ~5000 square feet of space dedicated to housing experimental animals under the direction of the Animal Resource Center in the Toxicology Department. The facility includes animal rooms as well as adjacent support rooms containing cage washers, sterilizers, storage areas, *etc.* Discussions with MBD staff and Animal Care Technicians revealed several P2Os dealing with sanitary waste and process sewer disposal. These are addressed in the "Pollution Prevention Opportunity Descriptions" numbers five and six.

Chemical Procurement

A second issue identified by the team was related to the procurement of chemicals. All eleven investigators within MBD procure their own chemicals on an as-needed basis from a number of vendors. Discussion with the staff indicated that it was common practice to determine whether the chemical was already available on site through informal questioning or the use of CHEM.FON. However, it was noted that orders were often placed without checking on internal availability. A possible P2O was identified and is addressed in the "Pollution Prevention Opportunity Description" number eight.

Business Policies

A final issue identified by the team was the effect of changes in business volume or product lines and the resulting changes in personnel on waste issues. It was noted that the extensive downsizing of the MBD during 1995 which included relatively sudden, involuntary lay-offs had resulted in the "generation" of waste. Specifically, the chemicals and reagents which were actively being used on a routine basis in canceled projects were unexpectedly placed in the waste stream. An operational dilemma which arose as a consequence was the need for remaining staff to identify and properly dispose of these materials. A possible P2O was identified and is addressed in the "Pollution Prevention Opportunity Description" number nine.

POLLUTION PREVENTION OPPORTUNITY ASSESSMENT
WORKSHEET 3
POLLUTION PREVENTION OPPORTUNITY DESCRIPTION

Date: August 14, 1996 **P2OA Code:** PNNL-96-05 **Facility:** 331 Building & Annex

Activity: Protein Determination

P2O No.: 1 **P2O Title:** Reduction of Wastes from Protein Assays

Current Practice

At least three different groups within the MBD routinely measure protein levels in biological samples using either the method of Lowry or Bradford. Current sample throughput is 2000 Lowry's and 1000 Bradford's per year. The Lowry assay generates 4.0 mL hazardous waste/assay and requires ~2 minutes of technical time per sample for set-up and analysis; the Bradford assay generates 5.0 mL waste per sample and also requires ~2 minutes of technical time per sample.

Recommended Action

Purchase an automated "Microplate Reader" which would allow the performance of both assays on a microscale reducing the amount of hazardous waste for both to 0.2 mL and reducing handling time to 1 minute per sample.

Calculation of Waste Reduction and/or Energy Savings

The current waste stream from this activity is ~13 kg hazardous waste per year. After implementation, this would be reduced to ~0.6 kg/year (12.4 kg reduction of hazardous waste).

Calculation of Annual Cost Savings

Current:

Labor (101 hr x $85/hr)	$8,585
Disposal (13 kg x $50/kg)	650
Reagents	573
Total	$9,808

Proposed:

Labor (51 hr x $85/hr)	$4,335
Disposal (0.6 kg x $50/kg)	30
Reagents	136
Total	$4,501

Annual Cost Savings:
$9,808 - $4,501 = $5,307

Calculation of Implementation Cost and Payback
Implementation

Microplate Reader	$15,000
Labor and Training (8 hr x $85/hr)	680
Total	$15,680

Payback (Implementation Costs/Annual Cost Savings):
$15,680/$5,307 = 2.9

Note: The waste disposal rates and hourly rates provided in Worksheet 3 (pages 13–25) of this appendix are for demonstration purposes only. They do *not* represent the actual rates used by PNNL.

POLLUTION PREVENTION OPPORTUNITY ASSESSMENT
WORKSHEET 3
POLLUTION PREVENTION OPPORTUNITY DESCRIPTION

Date: August 14, 1996 **P2OA Code:** PNNL-96-05 **Facility:** 331 Building & Annex

Activity: Custom Oligonucleotide Synthesis

P2O No.: 2 **P2O Title:** Outsourcing of Custom Oligonucleotide Synthesis

Current Practice
Custom oligonucleotides are currently synthesized on an as-needed basis in MBD using an Applied Biosystems Model 392 DNA/RNA Synthesizer. The system is engineered to perform the reactions on a microscale and utilizes prepackaged reagents supplied by the manufacturer. The current usage is ~200 syntheses/yr. Each reaction requires 1 hour of labor and approximately $2 of reagents and generates approximately 0.05 kg of waste.

Recommended Action
Outsource the custom synthesis of oligonucleotides. Given practical considerations, approximately 95% of these syntheses could be outsourced without compromising project timing.

Calculation of Waste Reduction and/or Energy Savings
The current waste stream from this activity is ~10 kg hazardous waste per year. Implementation would result in the elimination of 95% of this waste stream (9.5 kg reduction of hazardous waste).

Calculation of Annual Cost Savings
Current:

Labor (190 hr x $85/hr)	$16,150
Disposal (9.5 kg x $50/kg)	475
Reagents	380
Total	$17,005

Proposed:

Outsourcing (190 syntheses x $50/synthesis)	$9,500

Annual Cost Savings:

$17,005 - $9,500 = $7,505

Calculation of Implementation Cost and Payback
There are no implementation costs.

POLLUTION PREVENTION OPPORTUNITY ASSESSMENT
WORKSHEET 3
POLLUTION PREVENTION OPPORTUNITY DESCRIPTION

Date: August 14, 1996 **P2OA Code:** PNNL-96-05 **Facility:** 331 Building & Annex

Activity: Southern Blotting

P2O No.: 3 **P2O Title:** Elementary Neutralization

Current Practice

The process involves resolving DNA by electrophoretic mobility in an agarose gel and subsequent processing of the gel to transfer the DNA onto nylon sheets. This waste stream consists of 0.4 *M* sodium hydroxide used to depurinate the DNA prior to transfer. Since the resulting solution exhibits a pH above regulated limits (~13.5), the generated waste is hazardous due only to the characteristic of corrosivity (D002). This activity generates approximately 8 L of corrosive hazardous waste annually.

Recommended Action

The recommended action is to perform elementary neutralization on the waste to adjust the pH be between 6.0 and 10.0 making it acceptable for process sewer disposal. Concentrations of an appropriate acid (HCl) could be used to neutralize the waste such that the resulting salt concentrations would be below regulatory limits (<10% wt/vol for NaCl). This generating activity would require the preparation of generator documentation (training, contingency, waste analysis plans, and operating procedures) and would also involve record keeping and reporting activities.

Calculation of Waste Reduction and/or Energy Savings

Assuming all wastes can be adequately neutralized, total waste reduction would be approximately 8 kg of liquid hazardous waste annually.

Calculation of Annual Cost Savings

Current:		
	Disposal (8 kg x $50/kg)	$400
Proposed:		
	Labor (1 hr x $85/hr)	$ 85

Annual Cost Savings:

$400 - $85 = $315

Calculation of Implementation Cost and Payback

Implementation Costs:

Documentation Labor (25 hr x $85/hr)	$2,125

Payback (Implementation Costs/Annual Cost Savings):
$2,125/$315 = 6.7

POLLUTION PREVENTION OPPORTUNITY ASSESSMENT
WORKSHEET 3
POLLUTION PREVENTION OPPORTUNITY DESCRIPTION

Date: August 14, 1996 **P2OA Code:** PNNL-96-05 **Facility:** 331 Building & Annex

Activity: Multiple processes within MBD

P2O No.: 4 **P2O Title:** Ethanol (Solvent) Extraction

Current Practice
A number of processes performed in MBD result in the production of waste ethanol. The ethanol ranges in concentration from absolute to ~70% in water and contains a number of trace contaminants such as sodium acetate. Current production is approximately 60 kg per year of which approximately 6 kg is water.

Recommended Action
Purchase, install, and use a solvent recovery system. The Toxicology Department currently employs a system suitable for extracting ethanol; the data used to derive the calculation of Annual Cost Savings and Implementation Costs were obtained from KM Gideon in that department and are based on her best estimates of procurement prices, *etc.*

Calculation of Waste Reduction and/or Energy Savings
Under ideal extraction conditions, the waste reduction would be approximately 54 kg per year (60 kg total minus 6 kg water).

Calculation of Annual Cost Savings
Current:

Disposal (60 kg x $50/kg)	$3,000
Ethanol purchase ($1/kg)	54
	$3,054

Proposed:

Labor (12 hr x $85/hr)	$1,020
Disposal of residual (6 kg x $50/kg)	300
Total	$1,320

Annual Cost Savings:
$3,054 - $1,320 = $1,734

Calculation of Implementation Cost and Payback

Implementation Costs:

Purchase	$10,000
Facilities modification (*e.g.*, ventilation)	10,000
Total	$20,000

Payback (Implementation Costs/Annual Cost Savings):
$20,000/$1,734 = 11.5

POLLUTION PREVENTION OPPORTUNITY ASSESSMENT
WORKSHEET 3
POLLUTION PREVENTION OPPORTUNITY DESCRIPTION

Date: August 14, 1996 **P2OA Code:** PNNL-96-05 **Facility:** 331 Building & Annex

Activity: Multiple processes within MBD

P2O No.: 5 **P2O Title:** Wash Down of Animal Room

Current Practice
The 331 Building has an Animal Care Facility that is primarily used by the MBD. The flooring in the rooms require a "wash-down" as part of the routine housekeeping. Current practice is to perform this manually with mops making efforts to minimize the use of disinfectants difficult.

Recommended Action
No action is recommended. A new, automatic wash-down system is being installed as part of the D-423 project. One of the objectives of this project is to construct a new animal facility on the 1st floor of the 331 Building.

Calculation of Waste Reduction and/or Energy Savings
N/A

Calculation of Annual Cost Savings
N/A

Calculation of Implementation Cost and Payback
N/A

POLLUTION PREVENTION OPPORTUNITY ASSESSMENT
WORKSHEET 3
POLLUTION PREVENTION OPPORTUNITY DESCRIPTION

Date: August 14, 1996 **P2OA Code:** PNNL-96-05 **Facility:** 331 Building & Annex

Activity: Multiple processes within MBD

P2O No.: 6 **P2O Title:** Procurement of Animal Bedding Containing Recycled Material

Current Practice

The 331 Building has an animal care facility that is primarily used by the MBD. The facility uses wood chips for bedding (substrate at the bottom of rodent cages). It was suggested that a possible recycled wood chip product could be utilized if a vendor could be identified.

Recommended Action

No action is recommended. Alison Wagner, supervisor for the Animal Resources Center, indicates that no vendor exists that carries a recycled wood chip bedding product.

Calculation of Waste Reduction and/or Energy Savings

N/A

Calculation of Annual Cost Savings

N/A

Calculation of Implementation Cost and Payback

N/A

POLLUTION PREVENTION OPPORTUNITY ASSESSMENT
WORKSHEET 3
POLLUTION PREVENTION OPPORTUNITY DESCRIPTION

Date: August 14, 1996 **P2OA Code:** PNNL-96-05 **Facility:** 331 Building & Annex

Activity: High Performance Liquid Chromatography (HPLC)

P2O No.: 7 **P2O Title:** Reduction of HPLC Waste

Current Practice

Current practices vary and depend upon the specific analyte. In one case, an eluent was prepared using a premixed buffer and ~4% methanol. The mixture was used in the HPLC then collected for disposal. This waste stream also contained radiolabeled samples and was disposed through WM. This experiment is complete and no waste is being generated at this time. In other cases the eluents were mixtures of acetonitrile and a buffered solution or methanol also in a buffered solution.

Recommended Action

These are small volume generators and do not appear to be amenable to solvent recovery or separation (See P2O No. 4, Ethanol (Solvent) Extraction for an analysis of a waste which would be characterized as high volume for MBD). No action is recommended.

Calculation of Waste Reduction and/or Energy Savings

N/A

Calculation of Annual Cost Savings

N/A

Calculation of Implementation Cost and Payback

N/A

POLLUTION PREVENTION OPPORTUNITY ASSESSMENT
WORKSHEET 3
POLLUTION PREVENTION OPPORTUNITY DESCRIPTION

Date: August 14, 1996 **P2OA Code:** PNNL-96-05 **Facility:** 331 Building & Annex

Activity: Multiple processes within MBD

P2O No.: 8 **P2O Title:** Chemical Procurement Policy

Current Practice
The process involves the procurement of chemicals by the MBD. Chemical orders are currently placed through the Waste Management Field Services Representative (WMFSR) who prepares stores orders or purchase requisitions through the Chemical Management System (CMS). Chemicals are also received through the WMFSR where they are bar coded and entered into the CMS as being received.

Recommended Action
The recommended action was to prepare a formal procurement policy which would require that existing chemical inventories be referenced on PopChem prior to ordering new chemicals. However, during the investigation of this opportunity, it was learned that the ES&H Directorate is currently working to incorporate this feature as part of the CMS. The upgraded CMS will automatically check existing inventories and will provide the user with this information before a new chemical is purchased. This would streamline the process by eliminating the need to check the PopChem reference. It is recommended that further investigation of this opportunity be deferred until the development of this feature is completed. In the interim, it is recommended that the WMFSR reference the PopChem system to the extent practical prior to ordering new chemicals.

Calculation of Waste Reduction and/or Energy Savings
N/A

Calculation of Annual Cost Savings
N/A

Calculation of Implementation Cost and Payback
N/A

23

POLLUTION PREVENTION OPPORTUNITY ASSESSMENT
WORKSHEET 3
POLLUTION PREVENTION OPPORTUNITY DESCRIPTION

Date: August 14, 1996 **P2OA Code:** PNNL-96-05 **Facility:** 331 Building & Annex

Activity: Multiple processes within MBD

P2O No.: 9 **P2O Title:** Staff Exit Procedure

Current Practice

When research scientists and project managers leave the laboratory (retirement, voluntary or involuntary reduction, *etc.*) they often take with them valuable process knowledge and key information concerning hazardous materials and wastes they leave behind. Moreover, it is often the case that reagents which were in routine use during a staff member's work, even those which would not be characterized as hazardous, enter the waste stream upon termination. While several costly examples of this problem have been exhibited recently in Buildings 3745 and 3746, LSL-II, and Building 331, the LSL-II case will be used for this analysis. The issue in this case was the requirement for individuals other than the generator to characterize a large number of items utilizing general process knowledge so as to insure proper disposal. In this case, 3 researchers were involved and all of the costs have been divided by three to obtain a "cost/staff" value. The analysis assumes that one staff will leave employment every year.

Recommended Action

The general issues relating to this P2O are currently addressed in the MBD Conduct of Operations manual relating to control of equipment and system status (Section 8.3, "Status of Equipment and Systems"). The recommended action is to develop a formalized exit procedure applicable to all staff members explicitly stating the requirements upon leaving the department. This action would insure that individuals with direct process knowledge properly characterize all items without the need for extensive interactions between MBD professional/technical staff and waste management.

Calculation of Waste Reduction and/or Energy Savings

There would be no actual reduction in waste. However, the tendency of the staff dealing with the LSL-II case was to place materials into the waste stream regardless of their potential use to others (See P2O #8). An estimate is ~1/3 of the "waste" (~35 kg total for all three staff members) could have been used by others or recycled.

24

Calculation of Annual Cost Savings

Current:

Labor	
WMFSR: (80 hrs x $33.50/hr)/3	$893.33
RCT: (4 hrs x $64.30/hr)/3	85.73
WM: (20 hrs x $38.20/hr)/3	255.00
MBD Staff: (20 hrs x $33.65/hr)/3	224.33
Analysis (15 samples x $1,500.00/sample)/3	7,500.00
Disposal (35 kg x $50/kg)/3	583.33
Total	$9,541.72

Proposed:

Labor	
WMFSR: (2 hr x $33.50/hr)	$67.00
RCT: (1 hr x $64.30/hr)	64.30
MBD Staff: (8 hrs x $33.65/hr)	269.20
Disposal (10 kg x $50/kg)	500.00
Total	$900.50

Annual Cost Savings:

$9,541.72 - 900.50 = $8,641.22

Calculation of Implementation Cost and Payback

Implementation Costs:

Labor (8 hr x $85.00/hr)	$680.00

Payback (Implementation Costs/Annual Cost Savings):

$680.00/$8,641.22 = 0.1

POLLUTION PREVENTION OPPORTUNITY ASSESSMENT
WORKSHEET 4
POLLUTION PREVENTION OPPORTUNITY SUMMARY

Date: August 14, 1996 **P2OA Code:** PNNL-96-05 **Facility:** 331 Building & Annex

Activity: Biological Research within the MBD

P2O No.	P2O Title	Waste Class Reduced	Annual Waste Reduction or Energy Savings	Estimated Annual Savings	Estimated Implement. Cost	Pay-back (years)
1	Reduction of Wastes from Protein Assays	Hazardous	12.4 kg	$5,307	$15,680	2.9
2	Outsourcing of Custom Oligonucleotide Synthesis	Hazardous	9.5 kg	$17,005	$0	Immed.
3	Elementary Neutralization	Hazardous	8 kg	$315	$2,125	6.7
4	Ethanol (Solvent) Extraction	Hazardous	54 kg	$1,734	$20,000	11.5
5	Wash Down of Animal Rooms	Process Sewer	N/A	N/A	N/A	N/A
6	Procurement of Animal Bedding Containing Recycled Material	Sanitary	N/A	N/A	N/A	N/A
7	Reduction of HPLC Waste	Hazardous	N/A	N/A	N/A	N/A
8	Chemical Procurement Policy	Hazardous	N/A	N/A	N/A	N/A
9	Staff Exit Procedure	Hazardous	~3 kg	$8,641	$680	0.1

Notes and Other Benefits: It is recommended that P2Os 1-4 and 8, 9 be implemented (see Worksheet 5 for discussion).

26

POLLUTION PREVENTION OPPORTUNITY ASSESSMENT
WORKSHEET 5
FINAL SUMMARY

Date: August 14, 1996 **P2OA Code:** PNNL-96-05 **Facility:** 331 Building & Annex

Activity: Biological Research within the MBD

Proposed Opportunities and Discussion:

Nine P2Os were identified as part of this effort and of these, four (#1-#4) were found to yield quantifiable reductions in the waste streams from MBD. The first of these (Reduction of Wastes from Protein Assays, P20 #1) has already been more fully evaluated and submitted for funding (documentation is in Appendix B). The second (Outsourcing of Custom Oligonucleotide Synthesis, P20 #2) is already in the process of being implemented at the Department level. Similarly, the third (Elementary Neutralization, P20 #3) is already being addressed at the Laboratory level and implementation would be relatively simple once the necessary SOPs are available. The last (Ethanol (Solvent) Extraction, P20 #4) would result in a relatively large reduction in waste but is only marginally cost effective due to the high start-up costs. Regardless, the team will recommend that this opportunity be considered for implementation. Of the remaining five, the last two (Chemical Procurement Policy and Staff Exit Procedure, P20s #8 & #9) are unusual in that they conclude that "administrative" changes *may* yield a reduction in the waste stream. The team had a difficult time with these for two reasons. First, there was little empirical data on which to quantify potential reductions in the waste stream and the cost impact of these changes. Second, the waste issues involve other concerns such as personnel policies which are not readily amenable to change based solely on pollution prevention considerations. However, the team does believe that these two P2Os address real issues and were within the scope of consideration.

An important observation regarding the MBD which emerged from this assessment is that for purely financial reasons, the department is already rigorously evaluating its operations and implementing changes which have the consequence of reducing waste streams. The P2O "Outsourcing of Custom Oligonucleotide Synthesis" is an example of this ongoing process. An example not mentioned in this report is a major capital investment made in FY1995. Funds were found to purchase a Molecular Dynamics FluorImager Model SI. This device allows for the use of non-isotopic methods for a variety of molecular biological techniques replacing the use of ^{32}P labeling and resulted in a major reduction in this waste. The team recognizes this effort and believes that this process should be formalized at some level.

Recommendations and Schedule for Implementation:

Implementing P2Os numbers 1-3 should be done immediately. Obtaining a solvent extraction system is warranted based on the reduction in volume and efforts should be started immediately to identify a funding source. The team has identified an extraction system which is currently unused and could possibly be obtained without cost. The team also believes that the P2Os number eight and nine should be implemented as soon as possible. For these two opportunities, the team recommends that an effort be made to modify the Division's Environmental Safety & Health Plan or the MBD Conduct of Operations manual to explicitly include these issues. For example, statements could be added reinforcing the need for staff to insure that upon their leaving the company, they are obligated under existing law as "waste generators" to properly dispose of excess reagents, *etc*. A final recommendation is that the department formalize its ongoing efforts to cut costs which have the additional consequence of reducing the waste streams.

Appendix A: Sample Staff Interrogatory

Pollution Prevention Opportunity Assessment
Staff Interrogatory
April, 1996

PI: ______________________________

Date: ______________________________

Question: Please identify processes leading to the generation of:

1. Any LLW

2. Any LLMW

Question: Please identify the three processes which generate the largest volumes of:

3. HW

4. Process Sewered Waste

Appendix B: Reduction of Laboratory Wastes from Protein Assays

PROJECT: Reduction of Laboratory Waste from Protein Assays

PROJECT MANAGER: Kevin Groch
Molecular Biosciences Department
Health Division
Pacific Northwest National Laboratory

PROJECT TYPE: Source Reduction

WASTE TYPE: Hazardous

NARRATIVE:

Current Practice: At least three different groups within the Molecular Biosciences Department routinely measure protein levels in biological samples using either the method of Lowry or Bradford. Current sample throughput is 2000 Lowry's and 1000 Bradford's per year. The Lowry assay generates 4.0 ml hazardous waste/assay and requires approximately 2 minutes of technical time per sample for set-up and analysis; the Bradford assay generates 5.0 ml hazardous waste per sample and also requires 2 minutes of technical time per sample.

Proposal: Purchase an automated "Microplate Reader" which would allow the performance of both assays on a microscale reducing the amount of hazardous waste for both to 0.2 ml and reducing handling time to 1 minute per sample.

RETURN ON INVESTMENT:

Annual Operating Costs Before Implementation (B):

Reagent Costs	
4.2 Lowry kits ($99.00/kit)	$415.80
Cost to Manage and Dispose of Hazardous Waste	
8 kg Lowry ($50.00/kg)	400.00
5 kg Bradford ($50.00/kg)	250.00
Labor	
67 hours ($85/hr)	5,695.00
34 hours ($85/hr)	2,890.00
Total	$9,808.30

Appendix B: Reduction of Laboratory Wastes from Protein Assays (continued)

Annual Operating Costs After Implementation (A):

Reagent Costs:	
1.0 Lowry kits ($99.00/kit)	$99.00
0.5 Bradford kits ($75.00/kit)	37.50
Cost to Manage and Dispose of Hazardous Waste:	
0.4 kg Lowry ($50.00/kg)	20.00
0.2 kg Bradford ($50.00/kg)	10.00
Labor:	
34 hours ($85/hr)	2,890.00
17 hours ($85/hr)	1,445.00
Total	$4,501.50

Initial Instrument Investment (C):

Model 3550-UV Plate Reader	$10,300.00
Computer, Monitor, and Accessories	4,73.00
Total	$15,037.00

Expensed Installation Cost (E):

Labor and Training (8 hours)	$680.00

Useful Life (L) = 10 years

Annual Cost Savings: B - A - (C + E)/L = $3,735.10

$9,808.30 - 4,501.50 - (15,037.00 + 680)/10

Percent Return on Investment: (((B - A) / (C + E)) - 1/10) X 100% = 23.8%

(((9,808.30 - 4,501.50) / (15,037.00 + 680.00)) - (1/10)) X 100%

Payback in Years: (L (C + E)) / (L (B - A) - (C + E)) = 4.2

(10 (15,037.00 + 680)) / (10 (9,808.30 - 4,501.50) - (15,037.00 + 680.00)

Appendix B: Reduction of Laboratory Wastes from Protein Assays (continued)

EQUIPMENT INCLUDED IN START-UP INVESTMENT:

Item	Vendor	Cost
Model 3550-UV Microplate Reader	Bio-Rad	$10,300.00
Microplate Manager III Software	Bio-Rad	2,125.00
8-Pette Adjustable Pipettor	Bio-Rad	565.00
Interface Cabel	Bio-Rad	99.00
Pwer Macintosh 7200/90	MacWarehouse	1,549.00
Apple Multiple Scan Color Display	MacWarehouse	399.00
Total		$15,037.00

Appendix C: Process Sewer Summary

Reagent and Composition			Average Weekly Amount	Maximum Weekly Amount
A buffer (pH 7.5)				
sodium chloride	17.532 g/L	1.7%		
tris[hydroxymethyl]aminomethane•HCl	15.76 g/L	1.5%		
polyoxyethylene sorbitan monolaurate	3 g/L	0.3%	11.0 L	19.0 L
Dialysis Buffer				
glycerol	100 g/L	10%		
sodium chloride	17.532 g/L	1.5%		
tris[hydroxymethyl]aminomethane•HCl	15.76 g/L	0.3%	0.2 L	0.5 L
Hemaline				
sodium chloride	8.12 g/L	(0.8%)		
potassium phosphate monobasic	0.26 g/L	(<0.1%)		
sodium phosphate dibaseic	2.35 g/L	(0.2%)		
potassium chloride	0.28 g/L	(<0.1%)		
disodium ethylenediaminetetraacetic acid	0.36 g/L	(<0.1%)		
lithium chloride	0.43 g/L	(<0.1%)		3.0 L
			2.0 L	
MOPS Buffer (pH 8.0)				
3-[N-morpholino]propanesulfonic acid	4.186 g/L	(0.4%)		
sodium acetate	0.6566 g/L	(>0.1%)		
disodium ethylenediaminetetraacetic acid	3.722 g/L	(>0.1%)		3.0 L
			0.9 L	
PNM Buffer (pH 7.0)				
sodium phosphate (dibasic)	25.2 g/L	(2.5%)		
sodium phosphate (monobasic)	0.8 g/L	(<0.1%)		
Carnation non-fat dry milk	50 g/L	(5.0%)		
noniodet-40	0.5 mL/L	(<0.1%)	1.0 L	2.0 L
SC Solution				
sucrose	102.68 g/L	(10.3%)		
calcium chloride	0.588 g/L	(>0.1%)	2.0 L	4.0 L
SSC Buffer (pH 8.0)				
sodium chloride	8.9 g/L	(0.9%)		
sodium citrate	4.4 g/L	(0.4%)	5.0 L	14.0 L
STE Buffer (pH 7.0)				
sodium chloride	58.44 g/L	(5.8%)		
disodium ethylenediaminetetraacetic acid	0.372 g/L	(>0.1%)		
Tris[hydroxymethyl]aminomethane	1.57 g/L	(0.2%)	5.0 L	5.0 L
TAE Buffer (pH 7.0)				
Tris[hydroxymethyl]aminomethane	4.8 g/L	(0.5%)		
sodium acetate	1.6 g/L	(0.2%)		
disodium ethylenediaminetetraacetic acid	0.7 g/L	(<0.1%)	9.7 L	32.5

Appendix C: Process Sewer Summary (continued)

Reagent and Composition			Average Weekly Amount	Maxi-mum Weekly Amount
TBE Buffer (pH 7.0)				
Tris[hydroxymethyl]aminomethane	10.8 g/L	(1.1%)		
sodium borate	11.4g/L	(1.1%)		
disodium ethylenediaminetetraacetic acid	0.8 g/L	(<0.1%)	14.0 L	34.0
TBS Buffer (pH 7.0)				
Tris[hydroxymethyl]aminomethane	1.43 g/L	(0.1%)		
sodium chloride	7.7 g/L	(>0.1%)		
polyoxyethylene sorbitan monolaurate	1.0 ml/L	(0.1%)	4.0 L	4.0 L
TE Buffer (pH 7.0)				
Tris[hydroxymethyl]aminomethane	1.57 g/L	(0.2%)		
disodium ethylenediaminetetraacetic acid	2.92 g/L	(0.3%)	2.0 L	2.0 L
TG Buffer (pH 7.0)				
sodium chloride	8.0 g/L	(0.8%)		
sodium phosphate (dibasic)	25.2 g/L	(2.5%)		
sodium phosphate (monobasic)	0.8 g/L	(<0.1%)		
triton X-100	1.0 mL/L	(0.1%)		
bovine serum albumin	2.0 g/L	(0.2%)	60.1 L	100.2 L
Tris-Glycine Buffer (pH 8.0)				
Tris[hydroxymethyl]aminomethane	3.0 g/L	(0.3%)		
glycine	18.8 g/L	(1.8%)		
sodium dodecyl sulfate	1.0 g/L	(0.1%)	16.0	40.0
Clorox				
---			47 mL	110 mL
HEPES Stock Solution (2.0 *M*)				
N-2-hydroxyethylpiperazine-N'-2-ethane-sulfonic acid, sodium salt	460.6 g/L	46.1%	2.0 L	2.0 L
Magnesium Chloride Solution (100 m*M*)				
magnesium chloride	20.3 g/L	(2.0%)	0.5 L	0.5 L
Potassium Chloride Solution (100 m*M*)				
potassium chloride	7.75 g/L	(0.8%)	0.5 L	0.5 L
Monopotassium Phosphate Stock Solution (0.2 *M*)				
monopotassium phosphate	27.2 g/L	(2.7%)	5.0 L	5.0 L
Dipotassium Phosphate Stock Solution (0.2 *M*)				
dipotassium phosphate	45.64 g/L	(4.6%)	5.0 L	5.0 L
Sodium Acetate Stock Solution (3.0 *M*)				
sodium acetate	408.3 g/L	(40.8%)	5.0 L	5.0 L

Appendix C: Process Sewer Summary (continued)

Reagent and Composition			Average Weekly Amount	Maximum Weekly Amount
Sodium Dibromoacetate Solution				
sodium dibromoacetate		>0.1%	1.0 L	1.0 L
Sodium Bromochloroacetate Solution				
sodium bromochloroacetate		>0.1%	1.0 L	1.0 L
Sodium Bromodichloroacetate Solution				
sodium bromodichloroacetate		>0.1%	5.0 L	5.0 L
Sodium Dichloroacetate Solution				
sodium dichloroacetate		>0.1%	9.0 L	9.0 L
Sodium Trichloroacetate Solution				
sodium trichloroacetate		>0.1%	9.0 L	9.0 L
Monosodium Phosphate Stock Solution (0.2 *M*)				
monosodium phosphate	27.6 g/L	(2.8%)	5.0 L	5.0 L
Disodium Phosphate Stock Solution (0.2 *M*)				
disodium phosphate	28.4 g/L	(2.8%)	5.0 L	5.0 L
Tris Stock Solution (1.0 *M*)				
Tris[hydroxymethyl]aminomethane	157.6 g/L	(15.8%)	5.0 L	5.0 L
Dulbecco's Modified Eagle Medium (pH 7.4)				
1993-4 GibcoBRL Catalog (#11885)	p. 1/94			
N-2-hydroxyethylpiperazine-N'-2-ethane sulfonic acid	6.0 g/L	(0.6%)	1.0 L	2.0 L
Dulbecco's Phosphate Buffered Saline (pH 7.4)				
1993-4 GibcoBRL Catalog (#14040)	p. 2/14		13.2 L	28.0 L
F-12 Nutrient Mixture (pH 7.4)				
1993-4 GibcoBRL Catalog (#11765)	p. 1/115			
N-2-hydroxyethylpiperazine-N'-2-ethane sulfonic acid	6.0 g/L	(0.6%)	3.0 L	5.0 L
Hank's Balanced Salt Solution (pH 7.4)				
1993-4 GibcoBRL Catalog (#24020)	p. 2/15		0.3 L	3.0 L
Luria Broth				
1993-4 GibcoBRL Catalog (#12795-027)	p. 10/5		3.7 L	8.5 L

Appendix C: Process Sewer Summary (continued)

Reagent and Composition			Average Weekly Amount	Maximum Weekly Amount
Minimum Essential Medium (pH 7.4)				
1993-4 GibcoBRL Catalog (#11090)	p. 1/105-6			
N-2-hydroxyethylpiperazine-N'-2-ethane	6.0 g/L			
sulfonic acid		(0.6%)	0.7 L	3.0 L
RPMI 1640 (pH 7.4)				
1993-4 GibcoBRL Catalog (#11835)	p. 1/116		1.0 L	4.0 L
calf serum				
---			0.52 L	1.3 L
gentamycin solution				
1993-4 GibcoBRL Catalog (#15710-015)	p. 4/21		17.5 mL	35.0 mL
penicillin/streptomycin solution				
1993-4 GibcoBRL Catalog (#15070-014)	p. 4/23		10.0 mL	20.0 mL
collagenase/F-12 solution (pH 7.4)				
1993-4 GibcoBRL Catalog (#11765)	p. 1/115			
N-2-hydroxyethylpiperazine-N'-2-ethane				
sulfonic acid	6.0 g/L			
collagenase	1.0 g/L		0.1 L	0.5 L
trypsin-EDTA solution (pH 7.4)				
1993-4 GibcoBRL Catalog (#25200-015)	p. 4/40		.45 L	1.0 L

36

INDEX

A

B

C

D

E

F

H

I

K

L